# Digital Watermarking and Steganography: Fundamentals and Techniques (Second Edition)

Frank Y. Shih
New Jersey Institute of Technology

T0315164

CRC Press
Taylor & Francis Group
Boca Raton London New York

CRC Press is an imprint of the
Taylor & Francis Group, an **informa** business

CRC Press
Taylor & Francis Group
6000 Broken Sound Parkway NW, Suite 300
Boca Raton, FL 33487-2742

First issued in paperback 2020

© 2017 by Taylor & Francis Group, LLC
CRC Press is an imprint of Taylor & Francis Group, an Informa business

No claim to original U.S. Government works

ISBN-13: 978-1-4987-3876-7 (hbk)
ISBN-13: 978-0-367-65643-0 (pbk)

### Library of Congress Cataloging-in-Publication Data

Names: Shih, Frank Y., author.
Title: Digital watermarking and steganography : fundamentals and techniques / Frank Y. Shih.
Description: Second edition. | Boca Raton : Taylor & Francis, CRC Press, 2017. | Includes bibliographical references.
Identifiers: LCCN 2016048478| ISBN 9781498738767 (hardback : alk. paper) | ISBN 9781498738774 (ebook)
Subjects: LCSH: Digital watermarking. | Data encryption (Computer science) | Computer security. | Multimedia systems--Security measures. | Intellectual property.
Classification: LCC QA76.9.A25 S467 2017 | DDC 005.8/2--dc23
LC record available at https://lccn.loc.gov/2016048478

**Visit the Taylor & Francis Web site at**
**http://www.taylorandfrancis.com**

**and the CRC Press Web site at**
**http://www.crcpress.com**

# Dedication

*To my loving wife and children, to my parents who encouraged me through the years, and to those who helped me in the process of writing this book.*

# Contents

# Preface

Digital watermarking and steganography are important topics because digital multimedia is widely used and the Internet is rapidly growing. This book intends to provide a comprehensive overview of the different aspects, mechanisms, and techniques of information security. It is written for students, researchers, and professionals who take the related courses, want to improve their knowledge, and want to gain experience in digital watermarking and steganography.

Digital watermarking technology can be used to guarantee authenticity and as proof that the content has not been altered since insertion. Steganographic messages are often first encrypted by some traditional means, and then a covert text is modified in some way to contain the encrypted message. The need for information security exists everywhere, everyday.

This book aims to provide students, researchers, and professionals with technical information regarding digital watermarking and steganography, as well as instruct them on the fundamental theoretical framework for developing the extensive advanced techniques. By comprehensively considering the essential principles of the digital watermarking and steganographic systems, one can not only obtain novel ideas about implementing advanced algorithms but also discover new problems. The principles of digital watermarking and steganography in this book are illustrated with plentiful graphs and examples in order to simplify the problems, so readers can easily understand even complicated theories.

Several robust algorithms are presented in this book to illustrate the framework and to provide assistance and tools for understanding and implementing the fundamental principles. The combined spatial and frequency domain watermarking technique provides a new method of enlarging the embedding capacity of watermarks. The genetic algorithm (GA)-based watermarking technique solves the rounding error problem and is an efficient approach to embedding. The adjusted-purpose watermarking technique simplifies the selection of different types, and can be integrated into other watermarking techniques. The robust high-capacity watermarking technique successfully enlarges the hiding capacity while maintaining the watermark's robustness. GA-based steganography provides a new way of developing a robust steganographic system by artificially counterfeiting statistical features instead of using the traditional strategy of avoiding the alteration of statistical features.

## OVERVIEW OF THE BOOK

In Chapter 1, digital watermarking and digital steganography are briefly introduced. Then, the difference between watermarking and steganography is addressed. Next, a brief history along with updates on recently published resources is provided. The rest of the book is broken into two parts: Chapters 2 through 10 cover digital watermarking, and Chapters 11 and 12 cover digital steganography.

In Chapter 2, digital watermarking techniques are categorized, based on their characteristics, into five pairs: blind versus nonblind, perceptible versus imperceptible, private versus public, robust versus fragile, and spatial domain versus frequency domain. Digital watermarking techniques are classified, based on their applications, into five types: copyright protection, data authentication, fingerprinting, copy control, and device control. In Chapter 3, the basic mathematical preliminaries are introduced, including least-significant-bit substitution, the discrete Fourier transform, the discrete cosine transform, the discrete wavelet transform, random sequence generation, chaotic maps, error correction code, and set partitioning in hierarchical trees. In Chapter 4, the fundamentals of digital watermarking are introduced. The subject is divided into four classes: spatial domain, frequency domain, fragile, and robust. In the spatial domain class, substitutive watermarking and additive watermarking are introduced. In the frequency domain class, substitutive watermarking, multiplicative watermarking, vector quantization watermarking, and the rounding error problem are introduced. In the fragile watermark class, block-based watermarks, their weaknesses, and hierarchy-based watermarks are described. In the robust watermark class, the redundant embedding approach and spread spectrum are discussed.

In Chapter 5, the issue of watermarking attacks is explored. The attacks are summarized into four types: image-processing attacks, geometric attacks, cryptographic attacks, and protocol attacks. Then, they are further divided into four classes: filtering, remodulation, JPEG coding distortion, and JPEG 2000 compression. Next, geometric attacks are divided into nine classes: image scaling, rotation, image clipping, linear transformation, bending, warping, perspective projection, collage, and templates. After that, the cryptographic and protocol attacks are explained, and the available watermarking tools are provided. In the end, an efficient block-based fragile watermarking system is presented for tamper localization and the recovery of images. In Chapter 6, the technique of combinational digital watermarking is introduced. The combination of the spatial and frequency domains is described, and its advantages and experimental results are provided. The further encryption of combinational watermarks is explained. Chapter 7 shows how GAs can be applied to digital watermarking. The concept and basic operations of GAs are introduced and fitness functions are discussed. Then, GA-based rounding error correction watermarking is introduced. Next, the application of GA-based algorithms to medical image watermarking is presented. At the end of the chapter, the authentication of JPEG images based on GAs is described.

In Chapter 8, the technique of adjusted-purpose digital watermarking is introduced. Following an overview, the morphological approach to extracting pixel-based features, strategies for adjusting the variable-sized transform window (VSTW), and the quantity factor (QF) are presented. How VSTW is used to determine whether the embedded strategy should be in the spatial or the frequency domain and how the QF is used to choose fragile, semifragile, or robust watermarks are explained. An optimal watermarking solution is presented that considers imperceptibility, capacity, and robustness by using particle swarm optimization. Chapter 9 introduces a technique for robust high-capacity digital watermarking. After the weaknesses of current robust watermarking are pointed out, the concept of robust watermarking is introduced. Following this, the processes of enlarging the significant coefficients and breaking the local spatial similarity are explained. Next, new concepts are presented concerning block-based chaotic maps and the determination of embedding locations.

The design of intersection-based pixel collection, reference registers, and containers is described. A robust high-capacity watermarking algorithm and its embedding and extracting procedures are introduced. Experimental results are provided to explore capacity enlargement, robust experiments, and performance comparisons. At the end of the chapter, the technique of high-capacity multiple-regions-of-interest watermarking for medical images is presented. Chapter 10 introduces a novel fragile watermark-based reversible image authentication scheme. The chaotic hash value of each image block is computed as the watermark, which ensures that there is complicated nonlinear and sensitive dependence within the image's gray features, the secret key, and the watermark. Reversible watermark embedding allows the original image to be recovered exactly after the authentication message has been extracted. An improved reversible data-hiding algorithm using multiple scanning techniques and histogram modification is presented.

Chapters 11 and 12 cover the topic of digital steganography. In Chapter 11, three types of steganography are introduced: technical, linguistic, and digital. Some applications of steganography are illustrated, including convert communication and one-time pad communication. Concerns about embedding security and imperceptibility are explained. Four examples of steganography software are given: S-Tools, StegoDos, EzStego, and JSteg-Jpeg. Next, the concept of steganalysis, which intends to attack steganography, is discussed. The statistical properties of images, the visual steganalytic system (VSS), and the IQM-based steganalytic system are described. Three learning strategies—support vector machines, neural networks, and principle component analysis—are reviewed. At the end of the chapter, the frequency domain steganalytic system (FDSS) is presented. In Chapter 12, steganography based on GAs and differential evolution (DE) is introduced, which can break steganalytic systems. The emphasis is shifted from traditionally avoiding the alteration of statistical features to artificially counterfeiting them. An overview of GA-based breaking methodology is first presented. Then, GA-based breaking algorithms in the spatial domain steganalytic system (SDSS) are described. How one can generate stego-images in the VSS and in the IQM-based steganalytic system are explained. Next, the strategy of GA-based breaking algorithms in the FDSS is provided. Experimental results show that this algorithm can not only pass the detection of steganalytic systems but also increase the capacity of the embedded message and enhance the peak signal-to-noise ratio of stego-images. At the end of the chapter, we present the techniques of DE-based steganography. DE is a relative latecomer, but its popularity has been catching up. It is fast in numerical optimization and is more likely to find the true optimum.

## FEATURES OF THE BOOK

- New state-of-the-art techniques for digital watermarking and steganography
- Numerous practical examples
- A more intuitive development and a clear tutorial on the complex technology
- An updated bibliography
- Extensive discussion on watermarking and steganography
- The inclusion of steganalytic techniques and their counterexamples

## FEEDBACK ON THE BOOK

It is my hope that there will be opportunities to correct any errors in this book; therefore, please provide a clear description of any errors that you may find. Your suggestions on how to improve the book are always welcome. For this, please use either e-mail (shih@njit.edu) or regular mail to the author: Frank Y. Shih, College of Computing Sciences, New Jersey Institute of Technology, University Heights, Newark, NJ 07102-1982.

MATLAB® is a registered trademark of The MathWorks, Inc. For product information, please contact:

The MathWorks, Inc.
3 Apple Hill Drive
Natick, MA 01760-2098 USA
Tel: 508 647 7000
Fax: 508-647-7001
E-mail: info@mathworks.com
Web: www.mathworks.com

# Acknowledgments

Portions of the book appeared in earlier forms as conference papers, journal papers, or theses with my students here at the New Jersey Institute of Technology. Therefore, these parts of the text are sometimes a combination of my words and those of my students. I would like to gratefully acknowledge the Institute of Electrical and Electronic Engineers (IEEE) and Elsevier for giving me permission to reuse texts and figures that have appeared in some of my past publications.

**Frank Y. Shih**
*New Jersey Institute of Technology*

# Author

**Frank Y. Shih** received a BS from National Cheng-Kung University, Taiwan, in 1980, an MS from the State University of New York at Stony Brook in 1984, and a PhD from Purdue University, West Lafayette, Indiana, in 1987, all in electrical and computer engineering. He is jointly appointed as a professor in the Department of Computer Science, the Department of Electrical and Computer Engineering, and the Department of Biomedical Engineering at the New Jersey Institute of Technology, Newark. He currently serves as the director of the Computer Vision Laboratory.

Dr. Shih is currently on the editorial boards of the *International Journal of Pattern Recognition*, the *International Journal of Pattern Recognition Letters*, the *International Journal of Pattern Recognition and Artificial Intelligence*, the *International Journal of Recent Patents on Engineering*, the *International Journal of Recent Patents on Computer Science*, the *International Journal of Internet Protocol Technology*, and the *Journal of Internet Technology*. Dr. Shih has contributed as a steering member, committee member, and session chair for numerous professional conferences and workshops. He was the recipient of the Research Initiation Award from the National Science Foundation in 1991. He won the Honorable Mention Award for Outstanding Paper from the International Pattern Recognition Society and also won the Best Paper Award at the International Symposium on Multimedia Information Processing. He has received several awards for distinguished research at the New Jersey Institute of Technology. He has served several times on the Proposal Review Panel of the National Science Foundation.

Dr. Shih started his mathematical morphology research with applications to image processing, feature extraction, and object representation. His *IEEE Transactions on Pattern Analysis and Machine Intelligence (PAMI)* article "Threshold Decomposition of Grayscale Morphology into Binary Morphology" was a breakthrough solution to the bottleneck problem in grayscale morphological processing. His several articles in *IEEE Transactions on Image Processing* and *IEEE Transactions on Signal Processing* were innovations in fast exact Euclidean distance transformation and robust image enhancement and segmentation, using the *recursive soft morphological operators* he developed.

Dr. Shih further advanced the field of solar image processing and feature detection. In cooperation with physics researchers, he has made incredible contributions to bridge in the gap between solar physics and computer science. He and his colleagues have used these innovative computation and information technologies for real-time space weather monitoring and forecasting, and have received over $1 million in National Science Foundation grants. They have developed several methods to automatically detect and characterize filament/prominence eruptions, flares, and coronal mass ejections. These techniques are currently in use at the Big Bear Observatory in California as well as by NASA.

He has made significant contributions to mathematical morphology, pattern recognition, and information hiding, focusing on the security and robustness of digital watermarking and steganography. He has developed several novel methods to increase embedding capacity, enhance robustness, integrate different watermarking platforms, and break steganalytic systems. His recent article published in *IEEE Transactions on Systems, Man, and Cybernetics* is the first to apply GA-based methodology to breaking steganalytic systems.

Dr. Shih has so far published 130 journal papers, 100 conference papers, and 22 book chapters. The journals in which he has been published are top ranked in the professional societies. He has authored/edited four books: *Digital Watermarking and Steganography*, *Image Processing and Mathematical Morphology*, *Image Processing and Pattern Recognition*, and *Multimedia Security: Watermarking, Steganography, and Forensics*. He has overcome many difficult research problems in multimedia signal processing, pattern recognition, feature extraction, and information security. Some examples are robust information hiding, automatic solar feature classification, optimum feature reduction, fast accurate Euclidean distance transformation, and fully parallel thinning algorithms.

Dr. Shih is a research fellow for the American Biographical Institute and is a senior member of the IEEE. His current research interests include digital watermarking and steganography, digital forensics, image processing, computer vision, sensor networks, pattern recognition, bioinformatics, information security, robotics, fuzzy logic, and neural networks.

# 1 Introduction

Digital information and data are transmitted more often over the Internet now than ever before. The availability and efficiency of global computer networks for the communication of digital information and data have accelerated the popularity of digital media. Digital images, video, and audio have been revolutionized in the way they can be captured, stored, transmitted, and manipulated. This gives rise to a wide range of applications in education, entertainment, the media, industrial manufacturing, medicine, and the military, among other fields [1].

Computers and networking facilities are becoming less expensive and more widespread. Creative approaches to storing, accessing, and distributing data have generated many benefits for digital multimedia, mainly due to properties such as distortion-free transmission, compact storage, and easy editing. Unfortunately, free-access digital multimedia communication also provides virtually unprecedented opportunities to pirate copyrighted material. Therefore, the idea of using a digital watermark to detect and trace copyright violations has stimulated significant interest among engineers, scientists, lawyers, artists, and publishers, to name a few. As a result, research into the robustness of watermark embedding with respect to compression, image-processing operations, and cryptographic attacks has become very active in recent years, and the developed techniques have grown and been improved a great deal.

In this chapter, we introduce digital watermarking in Section 1.1 and digital steganography in Section 1.2. The differences between watermarking and steganography are given in Section 1.3. Finally, a brief history is described in Section 1.4.

## 1.1 DIGITAL WATERMARKING

Watermarking is not a new phenomenon. For nearly a thousand years, watermarks on paper have been used to visibly indicate a particular publisher and to discourage counterfeiting in currency. A watermark is a design impressed on a piece of paper during production and used for copyright identification (as illustrated in Figure 1.1). The design may be a pattern, a logo, or some other image. In the modern era, as most data and information are stored and communicated in digital form, proving authenticity plays an increasingly important role. As a result, digital watermarking is a process whereby arbitrary information is encoded into an image in such a way as to be imperceptible to observers.

Digital watermarking has been proposed as a suitable tool for identifying the source, creator, owner, distributor, or authorized consumer of a document or an image. It can also be used to detect a document or an image that has been illegally distributed or modified. Another technology, encryption, is the process of obscuring information to make it unreadable to observers without specific keys or knowledge. This technology is sometimes referred to as *data scrambling*. Watermarking, when

**FIGURE 1.1**   A paper watermark.

complemented by encryption, can serve a vast number of purposes including copyright protection, broadcast monitoring, and data authentication.

In the digital world, a watermark is a pattern of bits inserted into a digital medium that can identify the creator or authorized users. Digital watermarks—unlike traditional printed, visible watermarks—are designed to be invisible to viewers. The bits embedded into an image are scattered all around to avoid identification or modification. Therefore, a digital watermark must be robust enough to survive detection, compression, and other operations that might be applied to a document.

Figure 1.2 depicts a general digital watermarking system. A watermark message $W$ is embedded into a media message, which is defined as the host image $H$. The resulting image is the watermarked image $H^*$. In the embedding process, a secret key $K$—that is, a random number generator—is sometimes involved to generate a more secure watermark. The watermarked image $H^*$ is then transmitted along a communication channel. The watermark can later be detected or extracted by the recipient.

Imperceptibility, security, capacity, and robustness are among the many aspects of watermark design. The watermarked image must look indistinguishable from the original image; if a watermarking system distorts the host image to the point of being

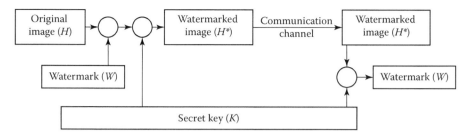

**FIGURE 1.2** A general digital watermarking system.

perceptible, it is of no use. An ideal watermarking system should embed a large amount of information perfectly securely, but with no visible degradation to the host image. The embedded watermark should be robust, with invariance to intentional (e.g., noise) or unintentional (e.g., image enhancement, cropping, resizing, or compression) attacks. Many researchers have focused on security and robustness, but rarely on watermarking capacity [2,3]. The amount of data an algorithm can embed in an image has implications for how the watermark can be applied. Indeed, both security and robustness are important because the embedded watermark is expected to be imperceptible and unremovable. Nevertheless, if a large watermark can be embedded into a host image, the process could be useful for many other applications.

Another scheme is the use of keys to generate random sequences during the embedding process. In this scheme, the cover image (i.e., the host image) is not needed during the watermark detection process. It is also a goal that the watermarking system utilizes an asymmetric key, as in public or private key cryptographic systems. A public key is used for image verification and a private key is needed for embedding security features. Knowledge of the public key neither helps compute the private key nor allows the removal of the watermark.

For user-embedding purposes, watermarks can be categorized into three types: *robust*, *semifragile*, and *fragile*. Robust watermarks are designed to withstand arbitrary, malicious attacks such as image scaling, bending, cropping, and lossy compression [4–7]. They are usually used for copyright protection in order to declare rightful ownership. Semifragile watermarks are designed for detecting any unauthorized modifications, while at the same time enabling some image-processing operations [8]. In other words, selective authentication detects illegitimate distortion while ignoring the applications of legitimate distortion. For the purpose of image authentication, fragile watermarks [9–13] are adopted to detect any unauthorized modification at all.

In general, we can embed watermarks in two types of domains: the spatial domain or the frequency domain [14–17]. In the spatial domain we can replace the pixels in the host image with the pixels in the watermark image [7,8]. Note that a sophisticated computer program may easily detect the inserted watermark. In the frequency domain, we can replace the coefficients of a transformed image with the pixels in the watermarked image [19,20]. The frequency domain transformations most commonly used are discrete cosine transform, discrete Fourier transform, and discrete wavelet transform. This kind of embedded watermark is, in general, difficult to detect.

However, its embedding capacity is usually low, since a large amount of data will distort the host image significantly. The watermark must be smaller than the host image; in general, the size of a watermark is one-sixteenth the size of the host image.

## 1.2  DIGITAL STEGANOGRAPHY

Digital steganography aims at hiding digital information in covert channels so that one can conceal the information and prevent the detection of the hidden message. Steganalysis is the art of discovering the existence of hidden information; as such, steganalytic systems are used to detect whether an image contains a hidden message. By analyzing the various features of stego-images (those containing hidden messages) and cover images (those containing no hidden messages), a steganalytic system is able to detect stego-images. Cryptography is the practice of scrambling a message into an obscured form to prevent others from understanding it, while steganography is the practice of obscuring the message so that it cannot be discovered.

Figure 1.3 depicts a classic steganographic model presented by Simmons [21]. In it, Alice and Bob are planning to escape from jail. All communications between them are monitored by the warden Wendy, so they must hide the messages in other innocuous-looking media (cover objects) in order to obtain each other's stego-objects. The stego-objects are then sent through public channels. Wendy is free to inspect all messages between Alice and Bob in one of two ways: passively or actively. The passive approach involves inspecting the message in order to determine whether it contains a hidden message and then to take proper action. The active approach involves always altering Alice's and Bob's messages even if Wendy may not perceive any traces of hidden meaning. Examples of the active method would be image-processing operations such as lossy compression, quality-factor alteration, format conversion, palette modification, and low-pass filtering.

For digital steganographic systems, the fundamental requirement is that the stego-image be perceptually indistinguishable to the degree that it does not raise suspicion. In other words, the hidden information introduces only slight modifications to the cover object. Most passive wardens detect the stego-images by analyzing their statistical features. In general, steganalytic systems can be categorized into two classes: spatial domain steganalytic systems (SDSSs) and frequency domain steganalytic systems (FDSSs). SDSSs [22,23] are adopted for checking lossless compressed images

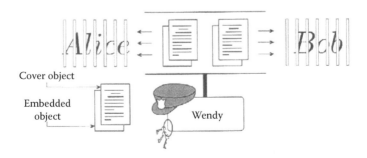

**FIGURE 1.3**   A classic steganographic model.

by analyzing the statistical features of the spatial domain. For lossy compressed images, such as JPEG files, FDSSs are used to analyze the statistical features of the frequency domain [24,25]. Westfeld and Pfitzmann have presented two SDSSs based on visual and chi-square attacks [23]. A visual attack uses human eyes to inspect stego-images by checking their lower bit planes, while a chi-square attack can automatically detect the specific characteristics generated by the least-significant-bit steganographic technique.

## 1.3  DIFFERENCES BETWEEN WATERMARKING AND STEGANOGRAPHY

Watermarking is closely related to steganography; however, there are some differences between the two. Watermarking mainly deals with image authentication, whereas steganography deals with hiding data. Embedded watermarking messages usually pertain to host image information such as copyright, so they are bound with the cover image. Watermarking is often used whenever the cover image is available to users who are aware of the existence of the hidden information and may intend to remove it. Hidden messages in steganography are usually not related to the host image. They are designed to make extremely important information imperceptible to any interceptors.

In watermarking, the embedded information is related to an attribute of the carrier and conveys additional information about or the properties of the carrier. The primary object of the communication channel is the carrier itself. In steganography, the embedded message usually has nothing to do with the carrier, which is simply used as a mechanism to pass the message. The object of the communication channel is the hidden message. As with the application of watermarking, a balance between image perceptual quality and robustness is maintained. Constraints in maintaining image quality tend to reduce the capacity of information embedded. As the application of steganography is different, dealing with covert message transfer, the embedded capacity is often viewed with as much importance as robustness and image quality.

## 1.4  A BRIEF HISTORY

The term *watermarking* is derived from the history of traditional papermaking. Wet fiber is pressed to expel the water, and the enhanced contrast between the watermarked and nonwatermarked areas of the paper forms a particular pattern and becomes visible.

Watermarking originated in the paper industry in the late Middle Ages—roughly, the thirteenth century. The earliest known usage appears to record the paper brand and the mill that produced it so that authenticity could be clearly recognized. Later, watermarking was used to certify the composition of paper. Nowadays, many countries watermark their paper, currencies, and postage stamps to make counterfeiting more difficult.

The digitization of our world has supplemented traditional watermarking with digital forms. While paper watermarks were originally used to differentiate between different manufacturers, today's digital watermarks have more widespread uses. Stemming from the legal need to protect the intellectual property of the creator from unauthorized usage, digital watermarking technology attempts to reinforce copyright by embedding a digital message that can identify the creator or the intended recipients. When encryption is broken, watermarking is essentially the technology to protect unencrypted multimedia content.

In 1989, Komatsu and Tominaga proposed digital watermarking to detect illegal copies [26]. They encoded a secret label into a copy using slight modifications to redundant information. When the label matches that of the registered owner, the provider can ensure that the document holder is the same person. As a method, digital watermarking has a long history, but it was only after 1990 that it gained large international interest. Today, a great number of conferences and workshops on this topic are held, and there are a large number of scientific journals on watermarking in publication. This renewed scientific interest in digital watermarking has quickly grabbed the attention of industry. Its widely used applications include copyright protection, labeling, monitoring, tamper proofing, and conditional access.

Watermarking or information embedding is a particular embodiment of steganography. The term *steganography* is derived from the Greek words for "covered or hidden" and "writing." It is intended to hide the information in a medium in such a manner that no one except the anticipated recipient knows the existence of the information. This is in contrast to cryptography, which focuses on making information unreadable to any unauthorized persons.

The history of steganography can be traced back to ancient Greece, where the hidden-message procedure included tattooing a shaved messenger's head, waiting for his hair to grow back, and then sending him out to deliver the message personally; the recipient would then shave the messenger's head once again in order to read the message. Another procedure included etching messages onto wooden tablets and covering them with wax. Various types of steganography and cryptography also thrived in ancient India, and in ancient China, military generals and diplomats hid secret messages on thin sheets of silk or paper. One famous story on the successful revolt of the Han Chinese against the Mongolians during the Yuan dynasty demonstrates a steganographic technique. During the Yuan dynasty (AD 1280–1368), China was ruled by the Mongolians. On the occasion of the Mid-Autumn Festival, the Han people made mooncakes (as cover objects) with a message detailing an attack plan inside (as hidden information). The mooncakes were distributed to members to inform them of the planned revolt, which successfully overthrew the Mongolian regime.

## REFERENCES

1. Berghel, H. and O'Gorman, L., Protecting ownership rights through digital watermarking, *IEEE Computer Mag.*, 29, 101, 1996.
2. Barni, M. et al., Capacity of the watermark channel: How many bits can be hidden within a digital image?, in *Proc. SPIE*, San Jose, CA, 1999, 437.

3. Shih, F. Y. and Wu, S.Y., Combinational image watermarking in the spatial and frequency domains, *Pattern Recognition*, 36, 969, 2003.

4. Cox, I., et al. Secure spread spectrum watermarking for images audio and video, in *Proc. IEEE Int. Conf. Image Processing*, Lausanne, Switzerland, 1996, 243.

5. Cox, I., et al. Secure spread spectrum watermarking for multimedia, *IEEE Trans. Image Processing*, 6, 1673, 1997.

6. Lin, S. D. and Chen, C.-F., A robust DCT-based watermarking for copyright protection, *IEEE Trans. Consumer Electronics*, 46, 415, 2000.

7. Nikolaidis, N. and Pitas, I., Robust image watermarking in the spatial domain, *Signal Processing*, 66, 385, 1998.

8. Acharya, U. R. et al., Compact storage of medical image with patient information, *IEEE Trans. Information Technology in Biomedicine*, 5, 320, 2001.

9. Caronni, G., Assuring ownership rights for digital images, in *Proc. Reliable IT Systems*, Vieweg, Germany, 1995, pp. 251–263.

10. Celik, M. et al., Hierarchical watermarking for secure image authentication with localization, *IEEE Trans. Image Processing*, 11, 585, 2002.

11. Pitas, I. and Kaskalis, T., Applying signatures on digital images, in *Proc. IEEE Workshop Nonlinear Signal and Image Processing*, Halkidiki, Greece, 1995, 460.

12. Wolfgang, R. and Delp, E., A watermarking technique for digital imagery: Further studies, in *Proc. Int. Conf. Imaging Science, Systems and Technology*, Las Vegas, NV, 1997.

13. Wong, P. W., A public key watermark for image verification and authentication, in *Proc. IEEE Int. Conf. Image Processing*, Chicago, IL, 1998, 425.

14. Langelaar, G. et al., Watermarking digital image and video data: A state-of-the-art overview, *IEEE Signal Processing Magazine*, 17, 20, 2000.

15. Cox, I. J. et al., Secure spread spectrum watermarking for multimedia, *IEEE Trans. Image Processing*, 6, 1673, 1997.

16. Petitcolas, F., Anderson, R., and Kuhn, M., Information hiding: A survey, *Proceedings of the IEEE*, 87, 1062, 1999.

17. Cox, I. and Miller M., The first 50 years of electronic watermarking, *J. Applied Signal Processing*, 2, 126, 2002.

18. Bruyndonckx, O., Quisquater, J.-J, and Macq, B., Spatial method for copyright labeling of digital images, in *Proc. IEEE Workshop Nonlinear Signal and Image Processing*, Neos Marmaras, Greece, 1995, 456.

19. Huang, J., Shi, Y. Q., and Shi, Y., Embedding image watermarks in DC components, *IEEE Trans. Circuits and Systems for Video Technology*, 10, 974, 2000.

20. Lin, S. D. and Chen, C.-F., A robust DCT-based watermarking for copyright protection, *IEEE Trans. Consumer Electronics*, 46, 415, 2000.

21. Simmons, G. J., Prisoners' problem and the subliminal channel, in *Proc. Int. Conf. Advances in Cryptology*, Santa Barbara, CA, 1984, 51.

22. Avcibas, I., Memon, N., and Sankur, B., Steganalysis using image quality metrics, *IEEE Trans. Image Processing*, 12, 221, 2003.

23. Westfeld, A. and Pfitzmann, A., Attacks on steganographic systems breaking the steganographic utilities EzStego, Jsteg, Steganos, and S-Tools and some lessons learned, in *Proc. Int. Workshop Information Hiding*, Dresden, Germany, 1999, 61.

24. Farid, H., Detecting steganographic messages in digital images, Technical Report, TR2001-412, Computer Science, Dartmouth College, 2001.

25. Fridrich, J., Goljan, M., and Hogea, D., New methodology for breaking steganographic techniques for JPEGs, in *Proc. SPIE*, Santa Clara, CA, 2003, 143.

26. Komatsu, N. and Tominaga H., A proposal on digital watermark in document image communication and its application to realizing a signature, *Trans. of the Institute of Electronics, Information and Communication Engineers*, J72B-I, 208, 1989.

## APPENDIX: SELECTED LIST OF BOOKS ON WATERMARKING AND STEGANOGRAPHY

Arnold, M., Wolthusen, S., and Schmucker, M., *Techniques and Applications of Digital Watermarking and Content Protection*, Norwood, MA: Artech House, 2003.

Baldoza, A., *Data Embedding for Covert Communications, Digital Watermarking, and Information Augmentation*, Washington, DC: Storming Media, 2000.

Barni, M. and Bartolini, F., *Watermarking Systems Engineering: Enabling Digital Assets Security and Other Applications*, Boca Raton, FL: CRC Press, 2004.

Chandramouli, R., *Digital Data-Hiding and Watermarking with Applications*, Boca Raton, FL: CRC Press, 2003.

Cole, E. and Krutz, R., *Hiding in Plain Sight: Steganography and the Art of Covert Communication*, New York: John Wiley, 2003.

Cox, I., Miller, M., and Bloom, J., *Digital Watermarking: Principles and Practice*, San Francisco: Morgan Kaufmann, 2001.

Cox, I., Miller, M., Bloom, J., Fridrich, J., and Kalker, T., *Digital Watermarking and Steganography*, 2nd edn., San Francisco: Morgan Kaufmann, 2007.

Cvejic, N. and Seppanen, T., *Digital Audio Watermarking Techniques and Technologies: Applications and Benchmarks*, New York: Information Science, 2007.

Eggers, J. and Girod, B., *Informed Watermarking*, New York: Springer, 2002.

Furht, B. and Kirovski, D., *Digital Watermarking for Digital Media*, New York: Information Science, 2005.

Furht, B. and Kirovski, D., *Fundamentals of Digital Image Watermarking*, New Jersey: John Wiley, 2005.

Furht, B. and Kirovski, D., *Multimedia Watermarking Techniques and Applications*, Boca Raton, FL: Auerbach, 2006.

Furht, B., Muharemagic, E., and, Socek, D., *Multimedia Encryption and Watermarking*, New York: Springer, 2005.

Gaurav, R., *Digital Encryption Model Using Embedded Watermarking Technique in ASIC Design*, Ann Arbor, MI: ProQuest/UMI, 2006.

Johnson, N., Duric, Z., and Jajodia, S., *Information Hiding: Steganography and Watermarking: Attacks and Countermeasures*, Norwell, MA: Kluwer Academic, 2001.

Katzenbeisser, S. and Petitcolas, F., *Information Hiding Techniques for Steganography and Digital Watermarking*, Norwell, MA: Artech House, 2000.

Kipper, G., *Investigator's Guide to Steganography*, Boca Raton, FL: CRC Press, 2003.

Kirovski, D., *Multimedia Watermarking Techniques and Applications*, Boca Raton, FL: Auerbach, 2006.

Kwok, S. et al., *Multimedia Security: Steganography and Digital Watermarking Techniques for Protection of Intellectual Property*, Hershey, PA: Idea Group, 2004.

Kwok, S., Yang, C., Tam, K., and Wong, J., *Intelligent Watermarking Techniques*, New Jersey: World Scientific, 2004.

Lu, C., *Multimedia Security: Steganography and Digital Watermarking Techniques for Protection of Intellectual Property*, Hershey, PA: Idea Group, 2005.

Pan, J., Huang, H., and Jain, L., *Intelligent Watermarking Techniques*, New Jersey: World Scientific, 2004.

Pfitzmann, A., *Information Hiding*, New York: Springer, 2000.

Seitz, J., *Digital Watermarking for Digital Media*, Hershey, PA: Idea Group, 2005.

Shih, F. Y., *Multimedia Security: Watermarking, Steganography, and Forensics,* Boca Raton, FL: CRC Press, 2013.

Su, J., *Digital Watermarking Explained*, New York: John Wiley, 2003.

Wayner, P., *Disappearing Cryptography: Information Hiding; Steganography and Watermarking*, 2nd edn., San Francisco: Morgan Kaufmann, 2002.

# 2 Classification in Digital Watermarking

With the rapidly increasing number of electronic commerce Web sites and applications, intellectual property protection is an extremely important concern for content owners who exhibit digital representations of photographs, books, manuscripts, and original artwork on the Internet. Moreover, as available computing power continues to rise, there is an increasing interest in protecting video files from alteration. Digital watermarking's applications are widely spread across electronic publishing, advertising, merchandise ordering and delivery, picture galleries, digital libraries, online newspapers and magazines, digital video and audio, personal communication, and more.

Digital watermarking is one of the technologies being developed as suitable tools for identifying the source, creator, owner, distributor, or authorized consumer of a document or image. It can also be used for tracing images that have been illegally distributed. There are two major steps in the digital watermarking process: (1) *watermark embedding*, in which a watermark is inserted into a host image, and (2) *watermark extraction*, in which the watermark is pulled out of the image. Because there are a great number of watermarking algorithms being developed, it is important to define some criteria for classifying them in order to understand how different schemes can be applied.

The chapter is organized as follows: Section 2.1 describes classification based on characteristics; Section 2.2 presents classification based on applications.

## 2.1 CLASSIFICATION BASED ON CHARACTERISTICS

This section categorizes digital watermarking technologies into five classes according to the characteristics of embedded watermarks.

1. Blind versus nonblind
2. Perceptible versus imperceptible
3. Private versus public
4. Robust versus fragile
5. Spatial domain based versus frequency domain based

### 2.1.1 BLIND VERSUS NONBLIND

A watermarking technique is said to be *blind* if it does not require access to the original unwatermarked data (image, video, audio, etc.) to recover the watermark. Conversely, a watermarking technique is said to be *nonblind* if the original data are needed for the extraction of the watermark. For example, the algorithms of Barni et al. and Nikolaidis and Pitas belong to the blind category, while those of Cox et al. and Swanson et al. belong to the nonblind [1–4]. In general, the nonblind scheme is more robust than the blind one because it is obvious that the watermark can be extracted easily by knowing the unwatermarked data. However, in most applications, the unmodified host signal is not available to the watermark detector. Since the blind scheme does not need the original data, it is more useful than the nonblind one in most applications.

### 2.1.2 Perceptible versus Imperceptible

A watermark is said to be *perceptible* if the embedded watermark is intended to be visible—for example, a logo inserted into a corner of an image. A good perceptible watermark must be difficult for an unauthorized person to remove and can resist falsification. Since it is relatively easy to embed a pattern or a logo into a host image, we must make sure the perceptible watermark was indeed the one inserted by the author. In contrast, an *imperceptible* watermark is embedded into a host image by sophisticated algorithms and is invisible to the naked eye. It could, however, be extracted by a computer.

### 2.1.3 Private versus Public

A watermark is said to be *private* if only authorized users can detect it. In other words, private watermarking techniques invest all efforts to make it impossible for unauthorized users to extract the watermark—for instance, by using a private, pseudorandom key. This private key indicates a watermark's location in the host image, allowing insertion and removal of the watermark if the secret location is known. In contrast, watermarking techniques that allow anyone to read the watermark are called *public*. Public watermarks are embedded in a location known to everyone, so the watermark detection software can easily extract the watermark by scanning the whole image. In general, private watermarking techniques are more robust than public ones, in which an attacker can easily remove or destroy the message once the embedded code is known.

There is also the *asymmetric* form of public watermarking, in which any user can read the watermark without being able to remove it. This is referred to as an *asymmetric cryptosystem*. In this case, the detection process (and in particular the detection key) is fully known to anyone, so only a public key is needed for verification and a private key is used for the embedding.

### 2.1.4 Robust versus Fragile

Watermark robustness accounts for the capability of the hidden watermark to survive legitimate daily usage or image-processing manipulation, such as intentional or unintentional attacks. Intentional attacks aim at destroying the watermark, while unintentional attacks do not explicitly intend to alter it. For embedding purposes, watermarks can be categorized into three types: (1) robust, (2) semifragile, and

(3) fragile. *Robust* watermarks are designed to survive intentional (malicious) and unintentional (nonmalicious) modifications of the image [2,3,5,6]. Unfriendly intentional modifications include unauthorized removal or alteration of the embedded watermark and unauthorized embedding of any other information. Unintentional modifications include image-processing operations such as scaling, cropping, filtering, and compression. Robust watermarks are usually used for copyright protection to declare rightful ownership.

*Semifragile* watermarks are designed for detecting any unauthorized modification, at the same time allowing some image-processing operations [7]. They are used for selective authentication that detects illegitimate distortion while ignoring applications of legitimate distortion. In other words, semifragile watermarking techniques can discriminate common image-processing and content-preserving noise—such as lossy compression, bit error, or "salt and pepper" noise—from malicious content modification.

For the purpose of authentication, *fragile* watermarks are adopted to detect any unauthorized modification [8–12]. Fragile watermarking techniques are concerned with complete integrity verification. The slightest modification of the watermarked image will alter or destroy the fragile watermark.

### 2.1.5 SPATIAL DOMAIN BASED VERSUS FREQUENCY DOMAIN BASED

There are two image domains for embedding watermarks: the spatial domain and the frequency domain. In the spatial domain [13], we can simply insert a watermark into a host image by changing the gray levels of some pixels in the host image. This has the advantages of low complexity and easy implementation, but the inserted information may be easily detected using computer analysis or could be easily attacked. We can embed the watermark into the coefficients of a transformed image in the frequency domain [3,14]. The transformations include discrete cosine transform, discrete Fourier transform, and discrete wavelet transform. However, if we embed too much data in the frequency domain, the image quality will be degraded significantly.

Spatial domain watermarking techniques are usually less robust to attacks such as compression and added noise. However, they have much lower computational complexity and usually can survive a cropping attack, which frequency domain watermarking techniques often fail to do. Another technique is combining both spatial domain watermarking and frequency domain watermarking for increased robustness and less complexity.

## 2.2 CLASSIFICATION BASED ON APPLICATIONS

Digital watermarking techniques embed hidden information directly into the media data. In addition to cryptographic schemes, watermarking represents an efficient technology to ensure data integrity as well as data origin authenticity. Copyright, authorized recipients, or integrity information can be embedded using a secret key into a host image as a transparent pattern. This section categorizes digital watermarking technologies into five classes according to the following applications:

1. Copyright protection
2. Data authentication
3. Fingerprinting
4. Copy control
5. Device control

### 2.2.1  COPYRIGHT PROTECTION

A watermark is invisibly inserted into an image that can be detected when the image is compared with the original. Watermarks for copyright protection are designed to identify both the source of the image as well as its authorized users. Public-key encryption, such as the Rivest, Shamir, and Adleman (RSA) algorithm [15], does not completely prevent unauthorized copying because of the ease with which images can be reproduced from previously published documents. All encrypted documents and images must be decrypted before the inspection. After the encryption is taken off, the document can be readable and disseminated. The idea of embedding an invisible watermark to identify ownership and recipients has attracted much interest in the printing and publishing industries.

Digital video can be copied repeatedly without loss of quality. Therefore, copyright protection for video data is more critical in digital video delivery networks than it is with analog TV broadcasting. One copyright protection method is to add a watermark to the video stream that carries information about the sender and recipient. In this way, video watermarking can enable the identification and tracing of different copies of video data. It can be applied to video distribution over the Internet, pay-per-view video broadcasting, and video disk labeling.

### 2.2.2  DATA AUTHENTICATION

The traditional means of data authentication are applied to a document in the form of a handwritten signature. Since the signature is affixed to a document, it is difficult to be modified or transported to another document. The comparison of two handwritten signatures can determine whether they were created by the same person.

A *digital signature* replicates the handwritten signature and offers an even stronger degree of authentication. A user can sign a digital document by encrypting it with a private key and an encryption scheme. Digital signatures make use of a public key or asymmetric cryptography, in which two keys are used that are related to each other mathematically. Users can verify digital signatures using only the public key but will need the secret key for the generation of digital signatures. Therefore, the public key is available to anyone who wishes to conduct verification, but the private key is merely given to authorized persons.

The data redundancy of an image makes it possible to authenticate the embedded watermark without accessing the original image. Although image format conversion leads to a different image representation, it does not change its visual appearance or authenticity. However, the authenticated image will inevitably be distorted by a small amount of noise due to the authentication itself. The distortion is often quite small, but this is usually unacceptable for medical imagery or images with a high strategic importance in military applications.

### 2.2.3 FINGERPRINTING

Biometrics technology, such as face, fingerprint, and iris recognition, plays an important role in today's personal identification systems. Digital document verification requires some reliable features such as fingerprinting. The digital watermarking of fingerprint images can be applied to protect the images against malicious attacks, detect fraudulent images, and ensure secure transmission. Fingerprinting in digital watermarking is usually used to embed an identity into an image in such a way that is difficult to erase. This allows the copyright owner to trace pirates if the image is distributed illegally. In this usage, watermarking is a method of embedding hidden information, known as the *payload*, within content. The content could be audio, image, or video, while the payload could identify the content owner or usage permission for the content. The payload of fingerprint watermarking is an identification number unique to each recipient of the content, the aim being to determine the source of illegally distributed copies.

### 2.2.4 COPY CONTROL

IBM's Tokyo Research Laboratory (TRL) first proposed the use of watermarking technology for DVD copy protection at the DVD Copy Protection Technical Working Group (CPTWG) in September 1996, and showed that the TRL's technology could detect embedded watermarks even in the MPEG2-compressed domain. This development enabled DVD playing and recording devices to automatically prevent the playback of unauthorized copies and unauthorized recording using copy control information detected in digital video content. The digital data are transmitted from a transmitter-side apparatus to a receiver-side apparatus through interfaces that allow the transmission and receiving of the digital data only between authenticated apparatuses. Copy control information indicating a copy restriction level is added to the main data recorded on the digital recording medium such that the main data comprises a first portion containing image and/or voice information and a second portion containing the copy control information. The digital watermark is embedded into the second portion of the main data.

### 2.2.5 DEVICE CONTROL

Device control watermarks are embedded to control access to a resource using a verifying device. A watermarking system embeds an authorization code into a signal and transmits it (e.g., as a television or radio program) to a verifying device. In the device, the authorization code is extracted from the watermarked signal and an operation to be performed on the resource is authorized separately on the extracted authorization code, which may consist of permission for executing a program or copying a multimedia object.

These watermarks can also be embedded in an audio signal to remotely control a device such as a toy, a computer, or an appliance. The device is equipped with an appropriate detector to identify hidden signals, which can trigger an action or change a state of the device. These watermarks can be used with a time gate device, whereby the watermark holds a time interval within which a user is permitted to conduct an action such as typing or pushing a button.

## REFERENCES

1. Barni, M. et al., A DCT-domain system for robust image watermarking, *Signal Processing*, 66, 357, 1998.
2. Nikolaidis, N. and Pitas, I., Robust image watermarking in the spatial domain, *Signal Processing*, 66, 385, 1998.
3. Cox, I. J. et al., Secure spread spectrum watermarking for multimedia, *IEEE Trans. Image Processing*, 6, 1673, 1997.
4. Swanson, M. D., Zhu, B., and Tewfik, A. H., Transparent robust image watermarking, in *Proc. IEEE Int. Conf. Image Processing*, Lausanne, Switzerland, 1996, 211.
5. Lin, S. D. and Chen, C.-F., A robust DCT-based watermarking for copyright protection, *IEEE Trans. Consumer Electronics*, 46, 415, 2000.
6. Deguillaume, F., Voloshynovskiy, S., and Pun, T., Secure hybrid robust watermarking resistant against tampering and copy attack, *Signal Processing*, 83, 2133, 2003.
7. Sun, Q. and Chang, S.-F., Semi-fragile image authentication using generic wavelet domain features and ECC, in *Proc. IEEE Int. Conf. Image Processing*, 2002, 901.
8. Wong, P. W., A public key watermark for image verification and authentication, in *Proc. IEEE Int. Conf. on Image Processing*, Chicago, IL, 1998, 425.
9. Celik, M. U. et al., Hierarchical watermarking for secure image authentication with localization, *IEEE Trans. Image Processing*, 11, 585, 2002.
10. Wolfgang, R. and Delp, E., A watermarking technique for digital imagery: Further studies, in *Proc. Int. Conf. Imaging Science, Systems and Technology*, Las Vegas, NV, 1997, 279.
11. Pitas, I. and Kaskalis, T., Applying signatures on digital images, in *Proc. IEEE Workshop Nonlinear Signal and Image Processing*, Neos Marmaras, Greece, 1995, 460.
12. Caronni, G., Assuring ownership rights for digital images, in *Proc. Int. Conf. Reliable IT Systems*, Vieweg, Germany, 1995, pp. 251–263.
13. Berghel, H. and O'Gorman, L., Protecting ownership rights through digital watermarking, *IEEE Computer Mag.*, 101, 1996.
14. Cox, I. et al., Secure spread spectrum watermarking for images audio and video, in *Proc. IEEE Int. Conf. Image Processing*, 1996, 243.
15. Chambers, W. G., *Basics of Communications and Coding*, Oxford Science, Clarendon Press, Oxford, 1985.

# 3 Mathematical Preliminaries

This chapter introduces the mathematical preliminaries of digital watermarking techniques for different embedding purposes and domains, and presents some commonly used operations in digital watermarking, including *least-significant-bit* (LSB) substitution, the *discrete Fourier transform* (DFT), the *discrete cosine transform* (DCT), the *discrete wavelet transform* (DWT), *random sequence generation*, *chaotic maps*, *error correction code* (ECC), and *set partitioning in hierarchical trees* (SPIHT).

## 3.1 LEAST-SIGNIFICANT-BIT SUBSTITUTION

LSB substitution is often used in image watermarking for embedding watermarks into a host image. In a grayscale image, a pixel is represented as eight bits with the *most significant bit* (MSB) to the left and the LSB to the right. For example, a pixel with a gray value of 130 is shown in Figure 3.1a. The idea of LSB substitution is to replace the LSB of a pixel with the watermark, because this has little effect on the appearance of the carrier message. For example, as shown in Figure 3.1b, when the LSB is changed, the pixel value changes from 130 to 131, which is undetectable in human perception. If we change bits other than the LSB, the image will be noticeably distorted. When we alter the bit closer to the MSB, the image will be distorted more. For instance, if we change the MSB, the pixel value 130 will be changed to 2, as shown in Figure 3.1c. This makes a significant change to the gray intensity. Sometimes, in order to embed more watermarks, the least number of bits are replaced. Since only the lower-order bits are altered, the resulting color shifts are typically imperceptible [1].

In general, the image formats adopted in LSB substitution are lossless. The LSB method is easy to implement and can possess high embedding capacity and low visual perceptibility because every cover bit contains one bit of the hidden message. However, because the data hidden in the LSB may be known, the LSB substitution method is impervious to watermark extraction and some attacks such as cropping and compression.

## 3.2 DISCRETE FOURIER TRANSFORM

In the early 1800s, French mathematician Joseph Fourier introduced the Fourier series for the representation of the continuous-time periodic signals. The signal can be decomposed into a linear weighted sum of harmonically related complex exponentials. This weighted sum represents the frequency content of a signal called the

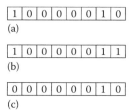

**FIGURE 3.1**   (a) An 8-bit pixel with a value of 130, (b) the value changed to 131 after the LSB substitution, (c) the value changed to 2 after the MSB substitution.

*spectrum.* When the signal becomes nonperiodic, its period becomes infinite and its spectrum becomes continuous. An image is considered a spatially varying function. The Fourier transform decomposes such an image function into a set of orthogonal functions and can transform the spatial intensity image into its frequency domain. From continuous form, one can obtain the form for discrete time images.

This section introduces the discrete case of a two-dimensional (2D) Fourier transform. If $f(x,y)$ denotes a digital image in the spatial domain and $F(u,y)$ denotes a transform image in the frequency domain, the general equation for a 2D DFT is defined as follows:

$$F(u,v) = \frac{1}{MN} \sum_{x=0}^{M-1} \sum_{y=0}^{N-1} f(x,y) \exp\left[-j2\pi\left(\frac{ux}{M} + \frac{vy}{N}\right)\right],$$   (3.1)

where $u = 0, 1, 2,\dots, M-1$ and $v = 0, 1, 2,\dots, N-1$. The inverse DFT can be represented as

$$f(x,y) = \sum_{u=0}^{M-1} \sum_{v=0}^{N-1} F(u,v) \exp\left[j2\pi\left(\frac{ux}{M} + \frac{vy}{N}\right)\right],$$   (3.2)

where $x = 0, 1, 2,\dots, M-1$ and $y = 0, 1, 2,\dots, N-1$. Because $F(u,v)$ and $f(x,y)$ are a Fourier transform pair, the grouping of the constant terms is not important. In practice, images are often sampled in a square array—that is, $M = N$. We use the following formulae:

$$F(u,v) = \frac{1}{N} \sum_{x=0}^{N-1} \sum_{y=0}^{N-1} f(x,y) \exp\left[-j2\pi\left(\frac{ux+vy}{N}\right)\right],$$   (3.3)

for $u,v = 0, 1, 2,\dots, N-1$, and

$$f(x,y) = \frac{1}{N} \sum_{u=0}^{N-1} \sum_{v=0}^{N-1} F(u,v) \exp\left[ j2\pi\left( \frac{ux+vy}{N} \right) \right], \qquad (3.4)$$

where $x,y = 0, 1, 2,\ldots, N-1$. The DFT can be used for phase modulation between the watermark image and its carrier, as well as for dividing the image into perceptual bands to record the watermark. The DFT uses phase modulation instead of magnitude components to hide messages since phase modulation has less visual effect. Furthermore, phase modulation is more robust against noise attack.

The number of complex multiplications and additions required to implement the DFT is proportional to $N^2$. Its calculation is usually performed using a method known as the *fast Fourier transform* [2], whose decomposition can make the number of multiplication and addition operations proportional to $N\log_2 N$.

## 3.3 DISCRETE COSINE TRANSFORM

The Fourier series was originally motivated by the problem of heat conduction and later found a vast number of applications as well as providing a basis for other transforms, such as the DCT. Many video and image compression algorithms apply the DCT to transform an image to the frequency domain and perform quantization for data compression. This helps separate an image into parts (or spectral sub-bands) of hierarchical importance (with respect to the image's visual quality). A well-known JPEG technology uses the DCT to compress images.

The Fourier transform kernel is complex valued. The DCT is obtained by using only a real part of the Fourier complex kernel. If $f(x,y)$ denotes an image in the spatial domain and $F(u,v)$ denotes an image in the frequency domain, the general equation for a 2D DCT is

$$F(u,v) = C(u)C(v) \sum_{x=0}^{N-1} \sum_{y=0}^{N-1} f(x,y) \cos\left( \frac{(2x+1)u\pi}{2N} \right) \cos\left( \frac{(2y+1)v\pi}{2N} \right), \qquad (3.5)$$

where if $u = v = 0$, $C(u) = C(v) = \sqrt{\frac{1}{N}}$ ; otherwise, $C(u) = C(v) = \sqrt{\frac{2}{N}}$.

The inverse DCT can be represented as

$$f(x,y) = \sum_{u=0}^{N-1} \sum_{v=0}^{N-1} C(u)C(v)F(u,v) \cos\left( \frac{(2x+1)u\pi}{2N} \right) \cos\left( \frac{(2y+1)v\pi}{2N} \right). \qquad (3.6)$$

A more convenient method for expressing the 2D DCT is with matrix products as $F = MfM^T$, and its inverse DCT is $f = M^T FM$, where $F$ and $f$ are $8 \times 8$ data matrices and $M$ is the following matrix:

$$
M = \begin{bmatrix}
\frac{1}{\sqrt{8}} & \frac{1}{\sqrt{8}} & \frac{1}{\sqrt{8}} & \frac{1}{\sqrt{8}} & \frac{1}{\sqrt{8}} & \frac{1}{\sqrt{8}} & \frac{1}{\sqrt{8}} & \frac{1}{\sqrt{8}} \\[2mm]
\frac{1}{2}\cos\frac{1}{16}\pi & \frac{1}{2}\cos\frac{3}{16}\pi & \frac{1}{2}\cos\frac{5}{16}\pi & \frac{1}{2}\cos\frac{7}{16}\pi & \frac{1}{2}\cos\frac{9}{16}\pi & \frac{1}{2}\cos\frac{11}{16}\pi & \frac{1}{2}\cos\frac{13}{16}\pi & \frac{1}{2}\cos\frac{15}{16}\pi \\[2mm]
\frac{1}{2}\cos\frac{2}{16}\pi & \frac{1}{2}\cos\frac{6}{16}\pi & \frac{1}{2}\cos\frac{10}{16}\pi & \frac{1}{2}\cos\frac{14}{16}\pi & \frac{1}{2}\cos\frac{18}{16}\pi & \frac{1}{2}\cos\frac{22}{16}\pi & \frac{1}{2}\cos\frac{26}{16}\pi & \frac{1}{2}\cos\frac{30}{16}\pi \\[2mm]
\frac{1}{2}\cos\frac{3}{16}\pi & \frac{1}{2}\cos\frac{9}{16}\pi & \frac{1}{2}\cos\frac{15}{16}\pi & \frac{1}{2}\cos\frac{21}{16}\pi & \frac{1}{2}\cos\frac{27}{16}\pi & \frac{1}{2}\cos\frac{33}{16}\pi & \frac{1}{2}\cos\frac{39}{16}\pi & \frac{1}{2}\cos\frac{45}{16}\pi \\[2mm]
\frac{1}{2}\cos\frac{4}{16}\pi & \frac{1}{2}\cos\frac{12}{16}\pi & \frac{1}{2}\cos\frac{20}{16}\pi & \frac{1}{2}\cos\frac{28}{16}\pi & \frac{1}{2}\cos\frac{36}{16}\pi & \frac{1}{2}\cos\frac{44}{16}\pi & \frac{1}{2}\cos\frac{52}{16}\pi & \frac{1}{2}\cos\frac{60}{16}\pi \\[2mm]
\frac{1}{2}\cos\frac{5}{16}\pi & \frac{1}{2}\cos\frac{15}{16}\pi & \frac{1}{2}\cos\frac{25}{16}\pi & \frac{1}{2}\cos\frac{35}{16}\pi & \frac{1}{2}\cos\frac{45}{16}\pi & \frac{1}{2}\cos\frac{55}{16}\pi & \frac{1}{2}\cos\frac{65}{16}\pi & \frac{1}{2}\cos\frac{75}{16}\pi \\[2mm]
\frac{1}{2}\cos\frac{6}{16}\pi & \frac{1}{2}\cos\frac{18}{16}\pi & \frac{1}{2}\cos\frac{30}{16}\pi & \frac{1}{2}\cos\frac{42}{16}\pi & \frac{1}{2}\cos\frac{54}{16}\pi & \frac{1}{2}\cos\frac{66}{16}\pi & \frac{1}{2}\cos\frac{78}{16}\pi & \frac{1}{2}\cos\frac{90}{16}\pi \\[2mm]
\frac{1}{2}\cos\frac{7}{16}\pi & \frac{1}{2}\cos\frac{21}{16}\pi & \frac{1}{2}\cos\frac{35}{16}\pi & \frac{1}{2}\cos\frac{49}{16}\pi & \frac{1}{2}\cos\frac{63}{16}\pi & \frac{1}{2}\cos\frac{77}{16}\pi & \frac{1}{2}\cos\frac{91}{16}\pi & \frac{1}{2}\cos\frac{105}{16}\pi
\end{bmatrix}.
$$

$$(3.7)$$

The obtained DCT coefficients indicate the correlation between the original $8 \times 8$ block and the respective DCT basis image. These coefficients represent the amplitudes of all cosine waves that are used to synthesize the original signal in the inverse process. The Fourier cosine transform inherits many properties from the Fourier transform, and there are other applications of the DCT that can be noted [3,4].

## 3.4  DISCRETE WAVELET TRANSFORM

The DWT is also a simple and fast transformation approach that translates an image from the spatial domain to the frequency domain. Unlike the DFT and DCT, which represent a signal either in the spatial or the frequency domain, the DWT is able to represent both interpretations simultaneously. It is used in JPEG 2000 compression and has become increasingly popular.

Wavelets are functions that integrate to zero waving above and below the $x$ axis. Like sines and cosines in the Fourier transform, wavelets are used as the basis functions for signal and image representation. Such base functions are obtained by dilating and translating a *mother wavelet* $\psi(x)$ by amounts $s$ and $\tau$, respectively.

$$
\Psi_{\tau,s}(x) = \left\{ \psi\left(\frac{x-\tau}{s}\right), (\tau,s) \in R \times R^{+} \right\}.
\tag{3.8}
$$

The translation and dilation allow the wavelet transform to be localized in time and frequency. Also, wavelet basis functions can represent functions with discontinuities and spikes in a more compact way than can sines and cosines.

The *continuous wavelet transform* (CWT) can be defined as

$$cwt_\psi(\tau, s) = \frac{1}{\sqrt{|s|}} \int x(t) \Psi_{\tau,s}^*(t) dt, \tag{3.9}$$

where $\Psi_{\tau,s}^*$ is the complex conjugate of $\Psi_{\tau,s}$ and $x(t)$ is the input signal defined in the time domain.

Inverse CWT can be defined as

$$x(t) = \frac{1}{C_\psi^2} \int_s \int_\tau cwt_\psi(\tau, s) \frac{1}{s^2} \Psi_{\tau,s}(t) d\tau ds, \tag{3.10}$$

where $C_\psi$ is a constant and depends on the wavelet used.

To discretize the CWT, the simplest case is the uniform sampling of the time–frequency plane. However, the sampling could be more efficient using the Nyquist rule.

$$N_2 = \frac{s_1}{s_2} N_1, \tag{3.11}$$

where $N_1$ and $N_2$ denote the number of samples at scales $s_1$ and $s_2$, respectively, and $s_2 > s_1$. This rule means that at higher scales (lower frequencies), the number of samples can be decreased. The sampling rate obtained is the minimum rate that allows the original signal to be reconstructed from a discrete set of samples. A dyadic scale satisfies the Nyquist rule by discretizing the scale parameter into a logarithmic series, and the time parameter is then discretized with respect to the corresponding scale parameters. The equations $\tau = k2^j$ and $s = 2^j$ set the translation and dilation, respectively, to the dyadic scale with a logarithmic series of base 2 for $\Psi_{k,j}$.

We can view these coefficients as filters that are classified into two types. One set, $H$, works as a low-pass filter, and the other, $G$, as a high-pass filter. These two types of coefficients are called *quadrature mirror filters* and are used in pyramidal algorithms.

For a 2D signal, the 2D wavelet transform can be decomposed by using a combination of one-dimensional (1D) wavelet transforms. The 1D transform can be applied individually to each of the dimensions of the image. By using quadrature mirror filters we can decompose an $n \times n$ image $I$ into the wavelet coefficients, as follows. Filters $H$ and $G$ are applied on the rows of an image, splitting the image into two subimages of dimensions $n/2 \times n$ (half the columns) each. One of these subimages, $H_r I$ (where the subscript $r$ denotes *row*), contains the low-pass information; the other, $G_r I$, contains the high-pass information. Next, the filters $H$ and $G$ are applied to the columns of both subimages. Finally, four subimages with dimensions $n/2 \times n/2$ are obtained. Subimages $H_c H_r I$, $G_c H_r I$, $H_c G_r I$, and $G_c G_r I$ (where the subscript $c$ denotes *column*) contain the low-low, high-low, low-high, and high-high passes, respectively. Figure 3.2 illustrates this decomposition. The same procedures are applied iteratively to the subimage containing the most low-band information until the subimage's size reaches $1 \times 1$. Therefore, the initial dimensions of the image are required to be powers of 2.

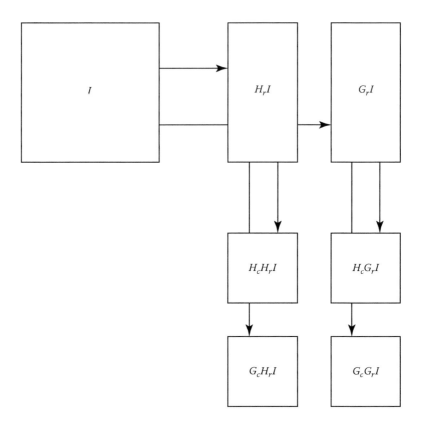

**FIGURE 3.2**    The wavelet decomposition of an image.

In practice, it is not necessary to carry out all the possible decompositions until the size $1 \times 1$ is reached. Usually, just a few levels are sufficient because most of the object features can be extracted from them. Figure 3.3 shows the *Lena* image decomposed in three levels. Each of the resulting subimages is known as a *sub-band*.

The Haar DWT [5,6] is the simplest of all wavelets (Figure 3.4). Haar wavelets are orthogonal and symmetric. The minimum support property allows arbitrary grid intervals and the boundary conditions are easier than those of other wavelet-based methods. The Haar DWT works very well to detect characteristics such as edges and corners. Figure 3.4 shows an example DWT of size $4 \times 4$.

## 3.5    RANDOM SEQUENCE GENERATION

After a watermark is embedded into a host image, if we crop or damage the image the recipient will not be able to extract the complete watermark. The extracted watermark will be distorted and some important information may be lost. To overcome the loss of information, we can rearrange the pixel sequence of a watermark via a random sequence generator [7].

For example, in Figure 3.5 we use the 14-bit random sequence generator to relocate the pixel order of a watermark of $128 \times 128$ by moving the 13th bit to the rear

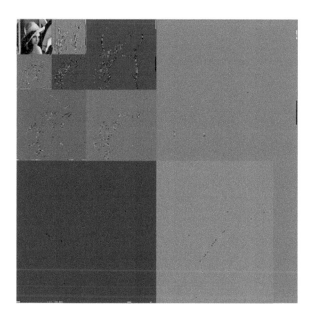

**FIGURE 3.3** The three-level wavelet decomposition of the *Lena* image.

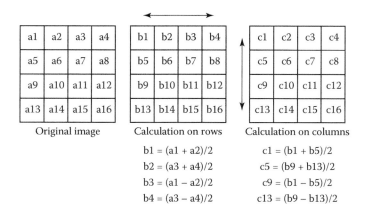

**FIGURE 3.4** An example of the Haar DWT.

(bit 0) of the whole sequence. A 2D watermark is first raster scanned into a 1D image of size 16,384 that can be stored in a 1D array with a 13-bit index. Let $(x,y)$ denote the column and row numbers respectively, and let the upper-left corner of the 2D image be (0,0). For example, a pixel located at (72,101) has the index 13000 (i.e., $128 \times 101 + 72$). Its binary code is 11001011001000 in the 1D array. After the 14-bit random sequence generator, the new index becomes 9617 (i.e., 10010110010001) in the 1D array. The corresponding location in the 2D image is (17,75), as shown in Figure 3.6.

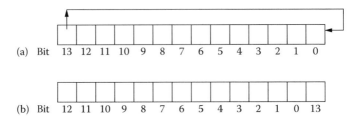

**FIGURE 3.5**    An example of random sequence generation.

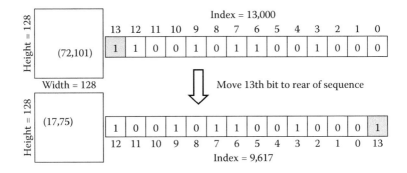

**FIGURE 3.6**    An example of pixel relocation by random sequence generation.

Another random sequence to spread the watermark is calculated using a secret key. Each value of the key is used as a seed to generate a short random sequence. When concatenating the short random sequences and randomly permuting the resulting sequence, we can obtain the pseudorandom sequence that will be used to spread the watermark. For example, the linear congruential generators are expressed as

$$I_{n+1} = (a \cdot I_n + b)(\bmod\ n), \tag{3.12}$$

where $a$, $b$, and $m$ are the chosen parameters. The initial value $I_0$ is randomly selected to generate the subsequent $I_n$ using those parameters. The initial input, called the *seed*, is usually short and is used to generate a long sequence of random numbers. If the same seed is used, the same output will be generated.

Another random sequence uses signature noise. Two keys, the owner key (the one known to the public) and the local key (the one calculated from image samples by a hash function), are used to generate the noise sequence. The local key and the owner key are combined to generate the random sequence, which results in a sequence with zero mean and an autocorrelation function close to an impulse. We can obtain a normal distribution by averaging each set of eight iterations of the generator. Eight have been selected because the division by a power of 2 is equivalent to a shift operator, which only consumes one central processing unit (CPU) cycle.

## 3.6   CHAOTIC MAP

Although the random sequence generation discussed in the previous section can rearrange the pixel permutation of a watermark, it does not spread the neighboring pixels into dispersed locations sufficiently. As a result, the relocated pixels are still close to one another. For example, Figure 3.7a shows an image containing a short line, and Figures 3.7b–e show the results after moving the 13th, 12th, 11th, and 10th bits to the rear location, respectively. They are still perceived as lines. This happens because the bit relocation approach does not change the spatial relationship among pixels. Table 3.1 illustrates a line of 11 pixels. We use random sequence generation to shift the 14th bit to the rear location. After pixel relocation, the distance between each pair of connected pixels is changed from 1 to 2. As an alternative, we can spread the pixels by swapping the 13th bit with the 0th bit. However, this will separate the pixels into two subgroups.

In order to spread the neighboring pixels into largely dispersed locations, we often use a chaotic map. In mathematics and physics, chaos theory deals with the behavior of certain nonlinear dynamic systems that exhibit a phenomenon under

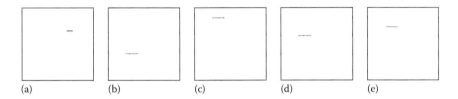

(a)                (b)                (c)                (d)                (e)

**FIGURE 3.7**   The pixel information in four different strategies of random sequence generation.

**TABLE 3.1**
**Pixel Information after Random Sequence Generation**

| Original Index | Binary Format | Original 2D Location | Move the 13th Bit to Rear | NewIndex | New 2D Location |
|---|---|---|---|---|---|
| 5200 | 01010001010000 | (80,40) | 10100010100000 | 10400 | (32,81) |
| 5201 | 01010001010001 | (81,40) | 10100010100010 | 10402 | (34,81) |
| 5202 | 01010001010010 | (82,40) | 10100010100100 | 10404 | (36,81) |
| 5203 | 01010001010011 | (83,40) | 10100010100110 | 10406 | (38,81) |
| 5204 | 01010001010100 | (84,40) | 10100010101000 | 10408 | (40,81) |
| 5205 | 01010001010101 | (85,40) | 10100010101010 | 10410 | (42,81) |
| 5206 | 01010001010110 | (86,40) | 10100010101100 | 10412 | (44,81) |
| 5207 | 01010001010111 | (87,40) | 10100010101110 | 10414 | (46,81) |
| 5208 | 01010001011000 | (88,40) | 10100010110000 | 10416 | (48,81) |
| 5209 | 01010001011001 | (89,40) | 10100010110010 | 10418 | (50,81) |
| 5210 | 01010001011010 | (90,40) | 1010001011010 | 10420 | (52,81) |

certain conditions known as *chaos* [8,9], which adopts the Shannon requirement on diffusion and confusion [10]. Its characteristics are used in robust digital watermarking to enhance security. The most critical aspects of chaotic functions are their utmost sensitivity to initial conditions and their outspreading behavior over the entire space. Therefore, chaotic maps are very useful in watermarking and encryption. The iterative values generated from chaotic functions are completely random in nature and never converge, although they are limited in range.

The logistic map is one of the simplest chaotic maps [11], expressed as

$$x_{n+1} = r \cdot x_n (1 - x_n), \tag{3.13}$$

where $x$ takes values in the interval $(0,1)$. If parameter $r$ is between 0 and 3, the logistic map will converge to a particular value after some iterations. As $r$ continues to increase, the curves' bifurcations become faster, and when $r$ is greater than 3.57, the curves complete chaotic periodicity. Finally, if $r$ is in the interval $(3.9,4)$, the chaotic values are in the range of $(0,1)$.

The chaotic map for changing pixel $(x,y)$ to $(x',y')$ can be represented by

$$\begin{bmatrix} x' \\ y' \end{bmatrix} = \begin{bmatrix} 1 & 1 \\ l & l+1 \end{bmatrix} \begin{bmatrix} x \\ y \end{bmatrix} \bmod N, \tag{3.14}$$

where $\det\left(\begin{bmatrix} 1 & 1 \\ l & l+1 \end{bmatrix}\right) = 1$ or $-1$, and $l$ and $N$ denote an integer and the width of a square image, respectively. The chaotic map is applied to the new image iteratively by Equation 3.14.

Figure 3.8 illustrates the chaotic map, as applied in Figure 3.8a to an image of size $101 \times 101$, which contains a short horizontal line. After using Equation 3.14 with $l = 2$ for four iterations, the results are shown in Figures 3.8b–e. We observe that the neighboring pixels are now scattered everywhere around the image. Table 3.2 lists the coordinate change of each pixel using the chaotic map.

Figure 3.9a shows a *Lena* image of size $101 \times 101$. Figures 3.9b1–17 are the iterative results of applying Equation 3.14 with $l = 2$. Note that the resulting image is exactly the same as the original image after 17 iterations.

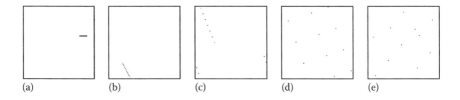

(a)   (b)   (c)   (d)   (e)

**FIGURE 3.8**   The pixel information in the first four iterations of a chaotic map.

**TABLE 3.2**
**The Coordinate Changes of Each Pixel during the Chaotic Map Procedure**

| Original 2D Location | 2D Location in First Iteration | 2D Location in Second Iteration | 2D Location in Third Iteration | 2D Location in Fourth Iteration |
|---|---|---|---|---|
| (80,40) | (19,78) | (97,70) | (66,0) | (66,31) |
| (81,40) | (20,80) | (100,78) | (77,30) | (6,42) |
| (82,40) | (21,82) | (2,86) | (88,60) | (47,53) |
| (83,40) | (22,84) | (5,94) | (99,90) | (88,64) |
| (84,40) | (23,86) | (8,1) | (9,19) | (27,75) |
| (85,40) | (24,88) | (11,9) | (20,49) | (69,86) |
| (86,40) | (25,90) | (14,17) | (31,79) | (9,97) |
| (87,40) | (26,92) | (17,25) | (42,8) | (50,7) |
| (88,40) | (27,94) | (20,33) | (53,38) | (91,18) |
| (89,40) | (28,96) | (23,41) | (64,68) | (31,29) |
| (90,40) | (29,98) | (26,49) | (75,98) | (72,40) |

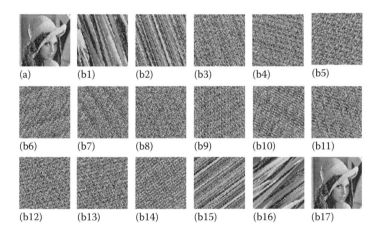

(a)    (b1)    (b2)    (b3)    (b4)    (b5)

(b6)    (b7)    (b8)    (b9)    (b10)    (b11)

(b12)    (b13)    (b14)    (b15)    (b16)    (b17)

**FIGURE 3.9**    An example of performing chaotic mapping on the *Lena* image.

## 3.7   ERROR CORRECTION CODE

ECC is another technology used to improve robustness in digital watermarking. It was designed to represent a sequence of numbers such that any errors occurring can be detected and corrected (within certain limitations) [12]. It is often used in coding theory to allow data that are being read or transmitted to be checked for errors and, when necessary, corrected. This is different from parity checking, which only detects errors. Nowadays, ECC is often designed into data storage and transmission hardware.

In digital watermarking, hidden data could be partially or completely destroyed due to some attacks, such as those from noise, smooth filters, and JPEG compression. In order to detect and correct the error message, specific coding rules and additional information are required. That is, after translating a watermark image into a bit-stream, we conduct reconstruction by specific rules and by adding some information, so a recipient is able to detect whether the bitstream (i.e., the watermark) is correct or incorrect. If the recipient detects that some bits of the bitstream are incorrect, he/she is able to correct them based on the additional information. Some commonly used ECCs are binary Golay code, Bose–Ray-Chaudhuri–Hocquenghem code, convolutional code, low-density parity check code, Hamming code, Reed–Solomon code, Reed–Muller code, and turbo code.

The basic coding approach is to translate the symbol into a binary code. Figure 3.10 shows a coding example in which 32 symbols (0–9 and A–V) are coded into a 5-bit binary stream. The information rate in Equation 3.15 is used to evaluate the coding efficiency, where $A$ is the code word and $n$ is the length of the coded bitstream:

$$\text{Information Rate} = \frac{\log_2 A}{n}. \tag{3.15}$$

The range of information rate is between 0 and 1. A higher value of information rate indicates a higher coding efficiency. For example, in Figure 3.10, there are 32 symbols ($A = 32$), and the length of the coded bitstream is 5 ($n = 5$). Therefore, the information rate is 1, meaning a perfect coding efficiency; that is, there is no waste in the coding.

If the watermarked image is under attack, the hidden data may be destroyed and the recipient will not be able to identify the correct message. For example, if a recipient extracts 00010 from a watermarked image, he/she cannot be sure whether this bitstream indicates 2 (if there is no bit error) or A (if the third bit was mistakenly changed from 1 to 0). We can provide specific coding rules, so the recipient can check the code to determine whether the message from the extracted watermark is altered. Figure 3.11 shows an example in which a recipient is able to know the correctness of the obtained message, where a 2-bit code containing four possibilities is

| 0 | 00000 | E | 01101 |
|---|-------|---|-------|
| 1 | 00001 | F | 01110 |
| 2 | 00010 | G | 01111 |
| ⋮ | ⋮ | ⋮ | ⋮ |
| 9 | 01001 | S | 11100 |
| A | 01010 | T | 11101 |
| B | 01011 | U | 11110 |
| C | 01100 | V | 11111 |

**FIGURE 3.10**    A coding example of 32 symbols.

| 2-bit coding | Receiver obtained |
|---|---|
| 4 cases: {00, 01, 10, 11} | 00   No error    00 |
| Only use two cases | 10 ⇒ Error ⇒ ?? |
| {00, ~~01, 10,~~ 11} | 01   Error      ?? |

**FIGURE 3.11**  An example of detecting the correctness of the obtained message.

given. The coding rule is defined as choosing only two cases, 00 and 11, for coding. If 01 or 10 is obtained, the recipient knows that the code is wrong. However, the recipient cannot be sure whether the correct message should be 00 or 11, since the minimum Hamming distance between the obtained message (10/01) and the correct code (00/11) is the same as 1. Note that the Hamming distance between two bitstreams of the same length is equal to the number of positions for which the corresponding symbols are different.

In order to correct the coding errors, additional information is required. For example, redundant codes are usually used to correct the errors. Figure 3.12 shows an example of the redundant code for correcting the error, where two copies of each bit are duplicated to form a 3-bit stream space. Therefore, a recipient can receive a total of eight cases within the 3-bit stream space. If the recipient obtains 001, 010, or 100, he/she knows there is a bit error and corrects it to be 000, so the message is 0. Similarly, if the recipient obtains 110, 101, or 011, he/she can correct it and determine that the message is 1.

*(7,4)Hamming* is a famous error-correcting code. For the remainder of this section, the Hamming code is introduced to check every four bits by adding three additional bits. The idea of the Hamming code is shown in Figure 3.13, where three additional bits are added and their values related to the original four bits. For example, bits 5, 6, and 7 are respectively derived by (bit 1 $\oplus$ bit 2 $\oplus$ bit 3), (bit 1 $\oplus$ bit 3 $\oplus$ bit 4), and (bit 2 $\oplus$ bit 3 $\oplus$ bit 4), where $\oplus$ denotes the *exclusive or* (XOR) Boolean operator. Therefore, a bitstream 1000 will become 1000110.

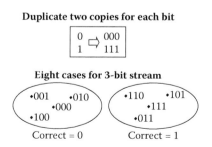

**Duplicate two copies for each bit**

0 ⇒ 000
1    111

**Eight cases for 3-bit stream**

•001    •010      •110    •101
   •000        •111
•100           •011

Correct = 0      Correct = 1

**FIGURE 3.12**  An example of redundant coding for error correction.

**Additional three bits**

$$a_1 \ a_2 \ a_3 \ a_4 \ | \ a_5 \ a_6 \ a_7 |$$

$$a_5 = a_1 \oplus a_2 \oplus a_3$$

$$a_6 = a_1 \oplus a_3 \oplus a_4$$

$$a_7 = a_2 \oplus a_3 \oplus a_4$$

| 1000 | 1000110 |
| 0100 | 0100101 |
| 0010 | 0010111 |
| 0001 | 0001011 |

**FIGURE 3.13**    The rule of adding 3 bits for Hamming code (7, 4).

Hamming matrices, as shown in Figure 3.14, including code generation matrices and parity check matrices, can be used to encode a 4-bit stream and detect errors of the obtained 7-bit coded stream. The method of encoding 16 possible Hamming codes by translating 4-bit streams into 7-bit coded streams is shown in Figure 3.15.

**Code generator matrix**          **Parity check matrix**

$$G = \begin{pmatrix} 1 & 0 & 0 & 0 & 1 & 1 & 0 \\ 0 & 1 & 0 & 0 & 1 & 0 & 1 \\ 0 & 0 & 1 & 0 & 1 & 1 & 1 \\ 0 & 0 & 0 & 1 & 0 & 1 & 1 \end{pmatrix} \qquad H = \begin{pmatrix} 1 & 1 & 0 \\ 1 & 0 & 1 \\ 1 & 1 & 1 \\ 0 & 1 & 1 \\ 1 & 0 & 0 \\ 0 & 1 & 0 \\ 0 & 0 & 1 \end{pmatrix}$$

**FIGURE 3.14**    The Hamming matrices.

**Method of encoding Hamming code**

$$(a_1 \ a_2 \ a_3 \ a_4) \begin{pmatrix} 1 & 0 & 0 & 0 & 1 & 1 & 0 \\ 0 & 1 & 0 & 0 & 1 & 0 & 1 \\ 0 & 0 & 1 & 0 & 1 & 1 & 1 \\ 0 & 0 & 0 & 1 & 0 & 1 & 1 \end{pmatrix} = (a_1 \ a_2 \ a_3 \ a_4 \ a_5 \ a_6 \ a_7)$$

**16 possible Hamming codes**

| 0000000 | 0001011 | 0010111 | 0011100 |
| 0100101 | 0101110 | 0110010 | 0111001 |
| 1000110 | 1001101 | 1010001 | 1011000 |
| 1100011 | 1101001 | 1110100 | 1111111 |

**FIGURE 3.15**    The 16 possible Hamming codes.

$$(1110100)\begin{pmatrix} 1 & 1 & 0 \\ 1 & 0 & 1 \\ 1 & 1 & 1 \\ 0 & 1 & 1 \\ 1 & 0 & 0 \\ 0 & 1 & 0 \\ 0 & 0 & 1 \end{pmatrix} = (000) \qquad (1011111)\begin{pmatrix} 1 & 1 & 0 \\ 1 & 0 & 1 \\ 1 & 1 & 1 \\ 0 & 1 & 1 \\ 1 & 0 & 0 \\ 0 & 1 & 0 \\ 0 & 0 & 1 \end{pmatrix} = (101)$$

(a)                                          (b)

**FIGURE 3.16**    An example of error correction based on the Hamming code.

An example of error correction is shown in Figure 3.16. By multiplying the obtained code stream and the parity check matrix and taking modulo-2, the syndrome vector can be obtained to indicate whether an error has occurred and, if so, for which code word bit. In Figure 3.16a, the syndrome vector is (0,0,0), indicating there is no error. If the original 7-bit stream was 1111111, then a single bit (i.e., the second bit) error occurred during transmission, causing the received 7-bit stream 1011111. In Figure 3.16b, the syndrome vector is (1,0,1), corresponding to the second row of the parity check matrix. Therefore, the recipient will know that the second bit of the obtained bitstream is incorrect.

## 3.8   SET PARTITIONING IN HIERARCHICAL TREE

SPIHT, introduced by Said and Pearlman [13], is a zerotree structure for image coding based on DWT. The first zerotree structure, called the *embedded zerotree wavelet* (EZW), was published by Shapiro in 1993 [14]. SPIHT coding uses a bit allocation strategy to produce a progressively embedded scalable bitstream.

A primary feature of SPIHT coding is that it converts the image into a sequence of bits based on the priority of the transformed coefficients of an image. The magnitude of each coefficient is taken to determine whether it is important. The larger the absolute value a coefficient has, the more important it is. Therefore, after SPIHT coding, the length of the coded string may be varied according to the quality requirement. In general, when a decoder reconstructs the image based on the coded string, a higher peak signal-to-noise ratio is achieved. The reconstructed image is similar to the original one if it can have the coded string as long as possible. There is no doubt that a decoder can reconstruct the original image if the completed coded string is available.

Basically, the SPIHT coding algorithm will encode the significant coefficients iteratively. In each iteration, if the absolute value of a coefficient is larger than the predefined threshold, the coefficient will be coded and its value (i.e., magnitude) will be decreased by a refinement procedure. The threshold is also

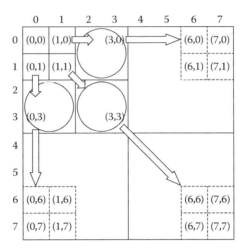

**FIGURE 3.17**   An example of quad-tree structure.

decreased in each iteration. The following are the relative terms in the SPIHT coding algorithm.

$n$: The largest factor determined based on the largest absolute coefficient by

$$n = \left\lfloor \log_2 \left[ \max \left( |c(i, j)| \right) \right] \right\rfloor. \tag{3.16}$$

$H$: A set of initial coefficients that are usually the low-frequency coefficients—for example, the coefficients (0,0), (0,1), (1,0), and (1,1) in Figure 3.17.

$LSP$: The list of significant partitioning coefficients. The significant coefficient determined in each iteration will be added into the LSP. It is set to be empty upon initialization.

$LIP$: The list of insignificant partitioning coefficients. Initially, all the coefficients in set $H$ are considered as insignificant and added into the LIP.

$LIS$: The list of the initial unprocessed coefficients.

$O(i,j)$: Set of all offspring of coefficient $(i,j)$. Excepting the highest and lowest pyramid levels in the quad-tree structure, $O(i,j)$ can be expressed by

$$O(i, j) = \left\{ (2i, 2j), (2i, 2j+1), (2i+1, 2j), (2i+1, 2j+1) \right\}. \tag{3.17}$$

For example, $O(3,3) = \{(6,6),(7,6),(6,7), \text{ and } (7,7)\}$ in Figure 3.17.

$D(i,j)$: The set of all descendants of coefficient $(i,j)$.

$L(i,j)$: The set of all descendants of coefficient $(i,j)$ except $D(i,j)$. That is, $L(i,j) = D(i,j) - O(i,j)$.

$c(i,j)$: The value of coefficient $(i,j)$ in the transformed image.

$S_n(i,j)$: The significant record of coefficient $(i,j)$. It is determined by

$$S_n(i, j) = \begin{cases} 1, & if \ |c(i, j)| \geq 2^n \\ 0, & \text{otherwise} \end{cases}.$$  (3.18)

*SPIHT coding algorithm*:

1. *Initialization*: Determine $n$ with Equation 3.15; set the LSP and LIS to be empty; add all elements in $H$ into the LIP.
2. *Sorting pass*:
   2.1 **For** each entry $(i,j)$ in the LIP **do**

   output $S_n(i, j)$;

   **if** $S_n(i, j) = 1$, **then** move $(i,j)$ into the LSP and out the sign of $c(i,j)$.

   2.2 **For** each entry $(i,j)$ in LIS **do**

   2.2.1 **If** the entry is of type A, **then**

   output $S_n(D(i, j))$;

   **if** $S_n(D(i, j)) = 1$, **then**

   **for** each $(k,l) \in O(i, j)$ **do**

   output $S_n(k,l)$;

   **if** $S_n(k,l) = 1$, **then** add $(k,l)$ into the LSP and out the sign of $c(k,l)$;

   **if** $S_n(k,l) = 0$, **then** add $(k,l)$ to the end of the LIP;

   **if** $L(i,j)! = 0$, **then** move $(i,j)$ to the end of the LIS, as an entry of type B;

   **else** remove entry $(i,j)$ from the LIS.

   2.2.2 **If** the entry is of type B, **then**

   output $S_n(L(i, j))$;

   **if** $S_n(L(i, j)) = 1$, **then**

   add each $(k,l) \in O(i, j)$ to the end of the LIS as an entry of type A;

   remove $(i,j)$ from the LIS.

3. *Refinement pass*: For each entry $(i,j)$ in the LSP, except those included in the last sorting pass, output the $n$th MSB bit of $c(i,j)$.
4. *Update*: Decrement $n$ by 1 and go to step 2.

Figure 3.18 shows an example of SPIHT, where Figure 3.18a is the original $8 \times 8$ image. We obtain the transformed image as Figure 3.18b by DWT. Figure 3.18c lists the bitstreams generated by applying SPIHT three times. We reconstruct the image by using these bitstreams, and the result is shown in Figure 3.18d.

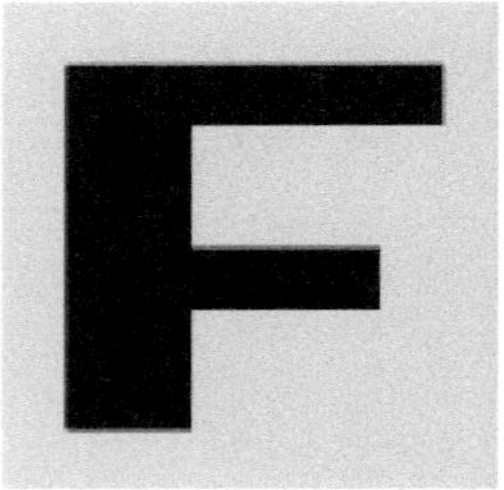

| 192 | 192 | 192 | 192 | 192 | 192 | 192 | 192 |
|-----|-----|-----|-----|-----|-----|-----|-----|
| 192 | 16  | 16  | 16  | 16  | 192 | 192 | 192 |
| 192 | 16  | 192 | 192 | 192 | 192 | 192 | 192 |
| 192 | 16  | 192 | 192 | 192 | 192 | 192 | 192 |
| 192 | 16  | 16  | 16  | 192 | 192 | 192 | 192 |
| 192 | 16  | 192 | 192 | 192 | 192 | 192 | 192 |
| 192 | 16  | 192 | 192 | 192 | 192 | 192 | 192 |
| 192 | 192 | 192 | 192 | 192 | 192 | 192 | 192 |

(a)

| 139.75 | -24.75 | 11  | -11 | 44  | 0   | 0   | -44 |
|--------|--------|-----|-----|-----|-----|-----|-----|
| -2.75  | 2.75   | 11  | -22 | 88  | -88 | 0   | 0   |
| 11     | -33    | 11  | -11 | 88  | -44 | 0   | 0   |
| -33    | -22    | 11  | -22 | 44  | -44 | 0   | 0   |
| 44     | 88     | 88  | 44  | -44 | 0   | 0   | 44  |
| 0      | 0      | -88 | 0   | 0   | 0   | 0   | 0   |
| 0      | -44    | 0   | 0   | 44  | 44  | 0   | 0   |
| -44    | -44    | 0   | 0   | 44  | -44 | 0   | 0   |

(b)

```
110000001
0000100001000001110011100001011001100001100010000001000000
0000010000101010010100000111111000101000101011010100101010011110111111001111111000000000000000000
```

(c)

| 138.25 | 0     | 0     | 0   | 40    | 0   | 0   | -40 |
|--------|-------|-------|-----|-------|-----|-----|-----|
| 0      | 0     | 0     | 0   | 92.5  | -92.5 | 0 | 0   |
| 0      | -40   | 0     | 0   | 92.5  | -40 | 0   | 0   |
| -40    | 0     | 0     | 0   | 40    | -40 | 0   | 0   |
| 40     | 92.5  | 92.5  | 40  | -40   | 0   | 0   | 40  |
| 0      | 0     | 0     | 0   | 0     | 0   | 0   | 0   |
| 0      | -40   | -92.5 | 0   | 40    | 40  | 0   | 0   |
| -40    | -40   | 0     | 0   | 40    | -40 | 0   | 0   |

(d)

**FIGURE 3.18**   An example of SPIHT: (a) Original 8 × 8 image, (b) image transformed by DWT, (c) the bitstreams by SPIHT, (d) the reconstructed image.

# REFERENCES

1. Johnson, N. and Jajodia, S., Exploring steganography: Seeing the unseen, *IEEE Computer*, 31, 26, 1998.
2. Brigham, E., *The Fast Fourier Transform and Its Applications*, Prentice-Hall, Englewood Cliffs, NJ, 1988.
3. Yip, P. and Rao, K., *Discrete Cosine Transform: Algorithms, Advantages, and Applications*, Academic Press, Boston, MA, 1990.
4. Lin, S. and Chen, C., A robust DCT-based watermarking for copyright protection, *IEEE Trans. Consumer Electronics*, 46, 415, 2000.
5. Falkowski, B. J., Forward and inverse transformations between Haar wavelet and arithmetic functions, *Electronics Letters*, 34, 1084, 1998.
6. Grochenig, K. and Madych, W. R., Multiresolution analysis, Haar bases, and self-similar tilings of $R^n$, *IEEE Trans. Information Theory*, 38, 556, 1992.
7. Knuth, D., *The Art of Computer Programming*, vol. 2, Addison-Wesley, Reading, MA, 1981.
8. Ott, E., *Chaos in Dynamical Systems*, Cambridge University Press, New York, 2002.
9. Zhao, D., Chen, G., and Liu, W., A chaos-based robust wavelet-domain watermarking algorithm, *Chaos, Solitons and Fractals*, 22, 47, 2004.
10. Schmitz, R., Use of chaotic dynamical systems in cryptography, *J. Franklin Institute*, 338, 429, 2001.
11. Devaney, R., *An Introduction to Chaotic Dynamical Systems*, 2nd edn., Addison-Wesley, Redwood City, CA, 1989.
12. Peterson, W. and Weldon, E., *Error-Correcting Codes*, MIT Press, Cambridge, MA, 1972.
13. Said, A. and Pearlman, W. A., A new, fast, and efficient image code based on set partitioning in hierarchical trees, *IEEE Trans. Circuits and Systems for Video Technology*, 6, 243, 1996.
14. Shapiro, J., Embedded image coding using zerotrees of wavelet coefficients, *IEEE Trans. Signal Processing*, 41, 3445, 1993.

# 4 Digital Watermarking Fundamentals

Digital watermarking is used for such information security as copyright protection, data authentication, broadcast monitoring, and covert communication. The watermark is embedded into a host image in such a way that the embedding-induced distortion is too small to be noticed. At the same time, the embedded watermark must be robust enough to withstand common degradations or deliberate attacks. Additionally, for given distortion and robustness levels, one would like to embed as much data as possible in a given host image. Podilchuk and Delp have presented a general framework for watermark embedding and detection/decoding [1], comparing the differences of various watermarking algorithms and applications in copyright protection, authentication, tamper detection, and data hiding.

This chapter introduces the fundamental digital watermarking techniques, including the spatial domain and frequency domain embedding approaches and fragile and robust watermarking techniques.

## 4.1 SPATIAL DOMAIN WATERMARKING

Spatial domain watermarking is the modifying of pixel values directly on the spatial domain of an image [2]. In general, spatial domain watermarking schemes are simple and do not need the original image to extract the watermark. They also provide a better compromise between robustness, capacity, and imperceptibility. However, they have the disadvantage of not being robust against image-processing operations because the embedded watermark is not distributed around the entire image and the operations can thus easily destroy the watermark. The purpose of watermarking is to embed a secret message at the content level under the constraints of imperceptibility, security, and robustness against attacks. We could categorize most watermark-embedding algorithms into substitution by a codebook element or additive embedding.

### 4.1.1 SUBSTITUTIVE WATERMARKING IN THE SPATIAL DOMAIN

Substitutive watermarking in the spatial domain is the simplest watermarking algorithm [3,4]. Basically, the embedding locations, such as the specific bits of all pixels, are predefined before watermark embedding. Once the recipient obtains the watermarked image, they know the exact locations from which to extract the watermark. During the watermark-embedding procedure, the watermark is first converted into a bitstream. Then, each bit of the bitstream is embedded into the specific bit of the

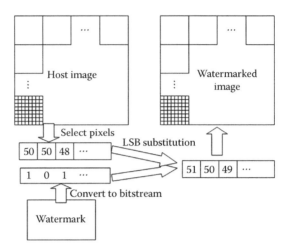

**FIGURE 4.1**  The watermark-embedding procedure of substitutive watermarking in the spatial domain.

selected locations for the host image. Figure 4.1 shows an example of substitutive watermarking in the spatial domain, in which there are three *least-significant-bit* (LSB) substitutions performed (i.e., embedding 1, 0, and 1 into 50, 50, and 48, respectively).

During the watermark extraction procedure, the specific pixel locations of the watermarked image are already known. Then, each pixel value is converted into its binary format. Finally, the watermark is collected from the bit in which it is embedded. Figure 4.2 shows an example of the watermark extracting procedure of substitutive watermarking in the spatial domain.

In general, the watermark capacity of substitutive watermarking in the spatial domain is larger than other watermarking approaches. Its maximum watermark capacity is eight times that of the host image. However, for the purpose of imperceptibility, a reasonable watermark capacity is three times that of the host image. If the watermark is embedded into the three LSBs, human beings cannot distinguish the original image from the watermarked image. Figure 4.3 shows an example of embedding a watermark into different bits. It is clear that the watermarked image can be distinguished by human eyes if the watermark is embedded into the seventh and fifth bits. However, if the watermark is embedded into the third bit, it would be difficult to tell whether there is a watermark hidden in the image.

Substitutive watermarking is simple to implement. However, the embedded watermark is not robust against collage or lossy compression attack. Therefore, some improved spatial domain watermarking algorithms have been proposed to target robustness—for example, the hash function method and the bipolar M-sequence method [5,6]. In contrast to robust watermarking, fragile watermarking can embed a breakable watermark into an image for the purpose of later detecting whether the image was altered.

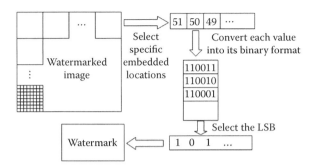

**FIGURE 4.2** The watermark extracting procedure of substitutive watermarking in the spatial domain.

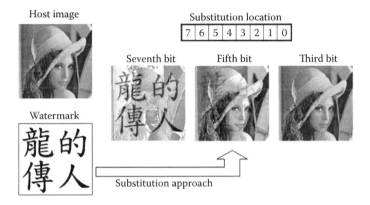

**FIGURE 4.3** An example of embedding a watermark into different bits.

## 4.1.2 ADDITIVE WATERMARKING IN THE SPATIAL DOMAIN

In contrast to the substitutive approach, the additive watermarking approach does not consider the specific bits of a pixel. Instead, it adds an amount of watermark value into a pixel to perform the embedding approach. If $H$ is the original grayscale host image and $W$ is the binary watermark image, then $\{h(i,j)\}$ and $\{w(i,j)\}$ denote their respective pixels. We can embed $W$ into $H$ to become the watermarked image $H^*$ as follows:

$$h*(i, j) = h(i, j) + a(i, j) \cdot w(i, j), \tag{4.1}$$

where $\{a(i,j)\}$ denotes the scaling factor. Basically, the larger $a(i,j)$ is, the more robust the watermarking algorithm is. Figure 4.4 shows an example of additive watermarking in the spatial domain. If the embedded watermark is 1, then $a(i,j) = 100$; otherwise, $a(i,j) = 0$. Therefore, after watermark embedding, the original values are changed from 50, 50, and 48 to 150, 50, and 148 using watermarks 1, 0, and 1, respectively.

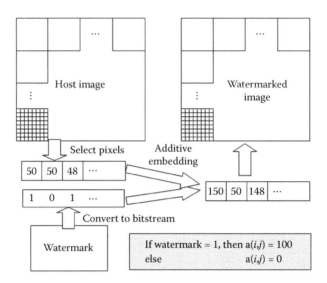

**FIGURE 4.4**    The watermark-embedding procedure of additive watermarking in the spatial domain.

It is obvious that if we have a large $a(i,j)$, the watermarked image would be distorted, as shown in Figure 4.3. It is then difficult to achieve high imperceptibility. Two critical issues to consider in the additive watermarking are imperceptibility and the need of the original image to identify the embedded message.

In order to enhance imperceptibility, a big value is not embedded to a single pixel, but to a block of pixels [7,8]. Figure 4.5 shows an example of the block-based additive watermarking. First, a 3 × 3 block is selected for embedding a watermark by adding 99. Second, the amount is divided by 9.

The second issue is the need for the original image in extracting the embedded watermark. When a recipient obtains the watermarked image, it is difficult to determine the embedded message since he does not know the location of the block for

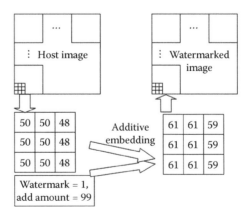

**FIGURE 4.5**    An example of block-based additive watermarking in the spatial domain.

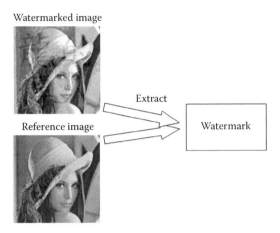

**FIGURE 4.6**    The reference image is required in the additive watermarking to extract the watermark.

embedding. Therefore, the reference (or original) image is usually required to extract the watermark, as shown in Figure 4.6.

## 4.2    FREQUENCY DOMAIN WATERMARKING

In frequency (or spectrum) domain watermarking [9,10], we can insert a watermark into frequency coefficients of the transformed image via the *discrete Fourier transform* (DFT), *discrete cosine transform* (DCT), or *discrete wavelet transform* (DWT). Because the frequency transforms usually decorrelate the spatial relationship of pixels, the majority of the energy concentrates on the low-frequency components. When we embed the watermark into the low or middle frequencies, these changes will be distributed throughout the entire image. When image-processing operations are applied to the watermarked image, they are less affected. Therefore, when compared with the spatial domain watermarking method, the frequency domain watermarking technique is relatively more robust.

This section describes the substitutive and multiplicative watermarking methods in the frequency domain, and will introduce a watermarking scheme based on *vector quantization* (VQ). Finally, it will present a rounding error problem in frequency domain methods.

### 4.2.1    SUBSTITUTIVE WATERMARKING IN THE FREQUENCY DOMAIN

The substitutive watermarking scheme in the frequency domain is basically similar to that of the spatial domain, except that the watermark is embedded into the frequency coefficients of the transformed image. Figure 4.7 shows an example of embedding a 4-bit watermark into the frequency domain of an image. Figure 4.7a is an 8 × 8 grayscale host image, and Figure 4.7b is the image transformed by DCT. Figure 4.7c is a binary watermark, in which 0 and 1 are the embedded values and a dash (–) indicates no change in position. We obtain Figure 4.7d by embedding

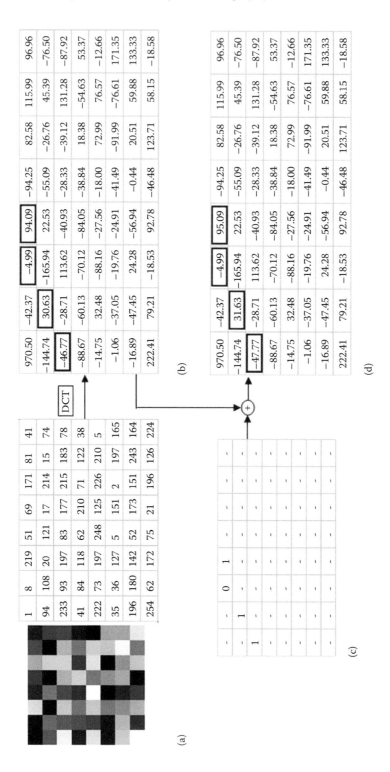

**FIGURE 4.7** An example of embedding a 4-bit watermark in the frequency domain of an image: (a) the original host image, (b) the transformed image, (c) the embedded watermark, (d) the transformed image after embedding watermark.

LSB substitution

| W | Original Coeff. | Integer part | Binary format | Watermarked | |
|---|---|---|---|---|---|
| | | | | Binary | Coeff. |
| 1 | −46.77 | 46 | 00101110 | 00101111 | −47.77 |
| 1 | 30.63 | 30 | 00011110 | 00011111 | 31.63 |
| 0 | −4.99 | 4 | 00000100 | 00000100 | −4.99 |
| 1 | 94.09 | 94 | 01011110 | 01011111 | 95.09 |

**FIGURE 4.8** Embedding a watermark into coefficients of the transformed image in Figure 4.7.

Figure 4.7c into Figure 4.7b using LSB substitution. Figure 4.8 shows the details of the LSB substitution approach.

## 4.2.2 MULTIPLICATIVE WATERMARKING IN THE FREQUENCY DOMAIN

Frequency domain embedding inserts the watermark into a prespecified range of frequencies in the transformed image. The watermark is usually embedded in the perceptually significant portion (i.e., the significant frequency component) of the image in order to be robust enough to resist attack. The watermark is scaled according to the magnitude of the particular frequency component. The watermark consists of a random, Gaussian-distributed sequence. This kind of embedding is called *multiplicative watermarking* [11]. If $H$ is the DCT coefficients of the host image and $W$ is the random vector, then $\{h(m,n)\}$ and $\{w(i)\}$ denote their respective pixels. We can embed $W$ into $H$ to become the watermarked image $H^*$ as follows:

$$h*(m,n) = h(m,n)(1 + \alpha(i) \cdot w(i)). \tag{4.2}$$

Note that a large $\{\alpha(i)\}$ would produce a higher distortion on the watermarked image. It is usually set to be 0.1 to provide a good trade-off between imperceptibility and robustness. An alternative embedding formula using the logarithm of the original coefficients is as follows:

$$h*(m,n) = h(m,n) \cdot e^{\alpha(i) \cdot w(i)}. \tag{4.3}$$

For example, let the watermark be a Gaussian sequence of 1000 pseudorandom real numbers. We select the 1000 largest coefficients in the DCT domain, and embed the watermark into a *Lena* image to obtain the watermarked image, as shown in Figure 4.9. The watermark can be extracted using the inverse embedding formula, as follows:

$$w'(i) = \frac{h*(m,n) - h(m,n)}{\alpha(i) \cdot h(m,n)}. \tag{4.4}$$

**FIGURE 4.9**    The watermarked image.

For a comparison of the extracted watermark sequence $\mathbf{w}'$ and the original watermark $\mathbf{w}$, we can use the similarity measure, as follows:

$$sim(\mathbf{w}', \mathbf{w}) = \frac{\mathbf{w}' \cdot \mathbf{w}}{|\mathbf{w}'|}. \tag{4.5}$$

The dot product of $\mathbf{w}' \cdot \mathbf{w}$ will be distributed according to the distribution of a linear combination of variables that are independent and normally distributed as follows:

$$N(0, \mathbf{w}' \cdot \mathbf{w}). \tag{4.6}$$

Thus, $sim(\mathbf{w}', \mathbf{w})$ is distributed according to $N(0,1)$. We can then apply the standard significance tests for the normal distribution. Figure 4.10 shows the response of the watermark detector to 500 randomly generated watermarks, of which only one matches the watermark. The positive response due to the correct watermark is much stronger than the response to incorrect watermarks [11]. If $\mathbf{w}'$ is created independently from $\mathbf{w}$, then it is extremely unlikely that $sim(\mathbf{w}', \mathbf{w}) > 6$. This suggests that the embedding technique has very low false-positive and false-negative response rates. By using Equation 4.5 on Figure 4.9, we obtain the similarity measure 29.8520.

### 4.2.3    WATERMARKING BASED ON VECTOR QUANTIZATION

A comprehensive review of VQ can be found in Gray [12]. It is a lossy block-based compression technique in which vectors, instead of scalars, are quantized. An image is partitioned into two-dimensional (2D) blocks. Each input block is mapped to a finite set of vectors forming the codebook that is shared by the encoder and the decoder. The index of the vector that best matches the input block based on some

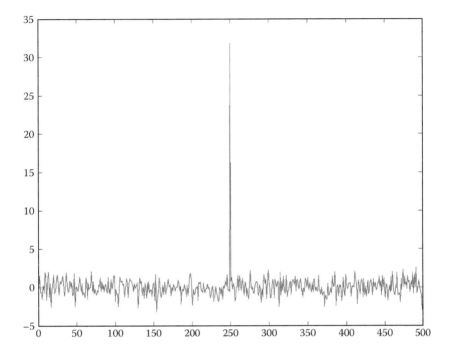

**FIGURE 4.10** Responses of 500 randomly generated watermarks, of which only one exactly matches the original watermark.

cost functions is sent to the decoder. While traditional VQ is a fixed-dimension scheme, a more general variable-dimension VQ has been developed.

A codebook is used to store all the vectors. Given an input vector **v**, Euclidean distance is used in the search process to measure the distance between two vectors, as follows:

$$k = \arg\min_j \sqrt{\sum_i (\mathbf{v}(i) - \mathbf{s}_j(i))^2}, \qquad (4.7)$$

where $j = 0, 1, \ldots, N-1$. The closest Euclidean distance code word $\mathbf{s}_k$ is selected and transmitted to the recipient. With the same codebook, the decomposition procedure can easily extract the vector **v** via table lookup.

Lu and Sun presented an image-watermarking technique based on VQ [13]. They divided the codebook into a few clusters of close code vectors with the Euclidean distance less than a given threshold. The scheme requires the original image for watermark extraction and uses the codebook partition as the secret key. When the received index is not in the same cluster as the best-match index, a tampering attack is detected.

## 4.2.4 ROUNDING ERROR PROBLEM

The rounding error problem exists in substitutive watermarking in the frequency domain. A recipient may be faced with the problem of correctly extracting the embedded watermark because the embedded message has been changed due to the modification of the coefficient values. Ideally, the data after performing the DCT or DWT, followed by an inverse transformation such as the *inverse discrete cosine transform* (IDCT) or *inverse discrete wavelet transform* (IDWT), should be exactly the same as the original data. However, if some data in the transformed image are changed, we will not obtain an image with all integers. This situation is illustrated in Figure 4.11a, which is the transformed image of Figure 4.7a. If we apply IDCT to Figure 4.11a, the original image can be reconstructed as Figure 4.11b. If the data enclosed by the bold rectangular boxes in Figure 4.11c are changed, the pixel values of the reconstructed image will not be integers, as in Figure 4.11d. Note that the changes between Figure 4.11a and Figure 4.11c are small.

Some researchers have suggested that the rounding technique be used to convert the real numbers to integers. Therefore, after applying the rounding approach to Figure 4.12a, we show the rounded image in Figure 4.12b. If the recipient wants to extract the watermark from the rounded image, they will perform a DCT to the rounded image and obtain its transformed image, as shown in Figure 4.12c. Unfortunately, the recipient cannot correctly extract the watermark from the location in which it is embedded even though there is only one difference, enclosed by the bold rectangle, between Figure 4.12b and its original image, Figure 4.7a.

Table 4.1 shows the results of embedding different kinds of watermarks into the same transformed image in Figure 4.7d. The embedding rule is shown in Figure 4.13. For a 4-bit watermark 1234, we insert 1, the most significant bit, into position A, 2 into position B, 3 into position C, and 4 into position D. There are $2^4$ possible embedding combinations. Among the 16 possibilities, only two cases of watermarks can be extracted correctly.

## 4.3 FRAGILE WATERMARKS

Due to the popularity of computerized image processing, there are a tremendous number of software tools available for users to employ in modifying images. Therefore, it becomes a critical problem for a recipient to judge whether the received image has been altered by an attacker. This situation is illustrated in Figure 4.14.

Fragile watermarks provide a means of ensuring that a received image can be trusted. The recipient will evaluate a hidden fragile watermark to see whether the image has been altered. Therefore, the fragile watermarking scheme is used to detect any unauthorized modification.

**(a)**

| | | | | | | | |
|---|---|---|---|---|---|---|---|
| 970.50 | -42.37 | -4.99 | 94.09 | -94.25 | 82.58 | 115.99 | 96.96 |
| -144.74 | 30.63 | -165.94 | 22.53 | -55.09 | -26.76 | 45.39 | -76.50 |
| -46.77 | -28.71 | 113.62 | -40.93 | -28.33 | -39.12 | 131.28 | -87.92 |
| -88.67 | -60.13 | -70.12 | -84.05 | -38.84 | 18.38 | -54.63 | 53.37 |
| -14.75 | 32.48 | -88.16 | -27.56 | -18.00 | 72.99 | 76.57 | -12.66 |
| -1.06 | -37.05 | -19.76 | -24.91 | -41.49 | -91.99 | -76.61 | 171.35 |
| -16.89 | -47.45 | 24.28 | -56.94 | -0.44 | 20.51 | 59.88 | 133.33 |
| 222.41 | 79.21 | -18.53 | 92.78 | -46.48 | 123.71 | 58.15 | -18.58 |

**(b)**

| | | | | | | | |
|---|---|---|---|---|---|---|---|
| 1 | 8 | 219 | 51 | 69 | 171 | 81 | 41 |
| 94 | 108 | 20 | 121 | 17 | 214 | 15 | 74 |
| 233 | 93 | 197 | 83 | 177 | 215 | 183 | 78 |
| 41 | 84 | 118 | 62 | 210 | 71 | 122 | 38 |
| 222 | 73 | 197 | 248 | 125 | 226 | 210 | 5 |
| 35 | 36 | 127 | 5 | 151 | 2 | 197 | 165 |
| 196 | 180 | 142 | 52 | 173 | 151 | 243 | 164 |
| 254 | 62 | 172 | 75 | 21 | 196 | 126 | 224 |

**(c)**

| | | | | | | | |
|---|---|---|---|---|---|---|---|
| 970.50 | -42.37 | -4.99 | 95.09 | -94.25 | 82.58 | 115.99 | 96.96 |
| -144.74 | 31.63 | -165.94 | 22.53 | -55.09 | -26.76 | 45.39 | -76.50 |
| -47.77 | -28.71 | 113.62 | -40.93 | -28.33 | -39.12 | 131.28 | -87.92 |
| -88.67 | -60.13 | -70.12 | -84.05 | -38.84 | 18.38 | -54.63 | 53.37 |
| -14.75 | 32.48 | -88.16 | -27.56 | -18.00 | 72.99 | 76.57 | -12.66 |
| -1.06 | -37.05 | -19.76 | -24.91 | -41.49 | -91.99 | -76.61 | 171.35 |
| -16.89 | -47.45 | 24.28 | -56.94 | -0.44 | 20.51 | 59.88 | 133.33 |
| 222.41 | 79.21 | -18.53 | 92.78 | -46.48 | 123.71 | 58.15 | -18.58 |

**(d)**

| | | | | | | | |
|---|---|---|---|---|---|---|---|
| 1.22 | 8.01 | 218.80 | 50.79 | 68.89 | 170.87 | 80.67 | 40.45 |
| 94.28 | 108.07 | 19.87 | 120.87 | 16.99 | 213.99 | 14.79 | 73.58 |
| 233.35 | 93.15 | 196.97 | 83.00 | 177.17 | 215.16 | 182.99 | 77.78 |
| 41.36 | 84.17 | 118.02 | 62.07 | 210.25 | 71.31 | 122.16 | 37.97 |
| 222.26 | 73.09 | 196.96 | 248.06 | 125.27 | 226.36 | 210.24 | 5.06 |
| 35.08 | 35.92 | 126.82 | 4.94 | 151.19 | 2.32 | 197.22 | 165.06 |
| 195.88 | 179.73 | 141.64 | 51.79 | 173.07 | 151.22 | 243.14 | 163.99 |
| 253.74 | 61.60 | 171.53 | 74.69 | 20.98 | 196.15 | 126.08 | 223.93 |

**FIGURE 4.11** An illustration that the pixel values in the reconstructed image will not be integers if some coefficient values are changed: (a) a transformed image, (b) the image after the IDCT from (a), (c) the transformed image after embedding the watermark, (d) the image after the IDCT from (c).

(a)

| 1.22 | 8.01 | 218.80 | 50.79 | 68.89 | 170.87 | 80.67 | 40.45 |
|---|---|---|---|---|---|---|---|
| 94.28 | 108.07 | 19.87 | 120.87 | 16.99 | 213.99 | 14.79 | 73.58 |
| 233.35 | 93.15 | 196.97 | 83.00 | 177.17 | 215.16 | 182.99 | 77.78 |
| 41.36 | 84.17 | 118.02 | 62.07 | 210.25 | 71.31 | 122.16 | 37.97 |
| 222.26 | 73.09 | 196.96 | 248.06 | 125.27 | 226.36 | 210.24 | 5.06 |
| 35.08 | 35.92 | 126.82 | 4.94 | 151.19 | 2.32 | 197.22 | 165.06 |
| 195.88 | 179.73 | 141.64 | 51.79 | 173.07 | 151.22 | 243.14 | 163.99 |
| 253.74 | 61.60 | 171.53 | 74.69 | 20.98 | 196.15 | 126.08 | 223.93 |

(b)

| 1 | 8 | 219 | 51 | 69 | 171 | 81 | 40 |
|---|---|---|---|---|---|---|---|
| 94 | 108 | 20 | 121 | 17 | 214 | 15 | 74 |
| 233 | 93 | 197 | 83 | 177 | 215 | 183 | 78 |
| 41 | 84 | 118 | 62 | 210 | 71 | 122 | 38 |
| 222 | 73 | 197 | 248 | 125 | 226 | 210 | 5 |
| 35 | 36 | 127 | 5 | 151 | 2 | 197 | 165 |
| 196 | 180 | 142 | 52 | 173 | 151 | 243 | 164 |
| 254 | 62 | 172 | 75 | 21 | 196 | 126 | 224 |

(c)

| 970.38 | -42.20 | -5.16 | 94.24 | -94.38 | 82.68 | 115.92 | 96.99 |
|---|---|---|---|---|---|---|---|
| -144.91 | 30.87 | -166.17 | 22.73 | -55.26 | -26.62 | 45.30 | -76.46 |
| -46.94 | -28.49 | 113.41 | -40.74 | -28.49 | -38.99 | 131.19 | -87.88 |
| -88.82 | -59.93 | -70.31 | -83.88 | -38.99 | 18.49 | -54.71 | 53.41 |
| -14.88 | 32.65 | -88.32 | -27.41 | -18.13 | 73.09 | 76.50 | -12.62 |
| -1.16 | -36.91 | -19.89 | -24.79 | -41.58 | -91.91 | -76.66 | 171.37 |
| -16.95 | -47.36 | 24.19 | -56.86 | -0.51 | 20.56 | 59.84 | 133.35 |
| 222.37 | 79.26 | -18.58 | 92.82 | -46.51 | 123.73 | 58.13 | -18.57 |

(d)

| - | - | 0 | - | - | - | - | - |
| - | 1 | - | - | - | - | - | - |
| - | 0 | - | - | - | - | - | - |
| - | 0 | - | | Original | | - | - |
| - | 0 | - | | watermark | | - | - |
| - | - | - | | 1101 | | - | - |
| - | - | - | - | - | - | - | - |
| - | - | - | - | - | - | - | - |

**FIGURE 4.12** An example showing that the watermark cannot be correctly extracted: (a) an image after the IDCT, (b) translating real numbers into integers by rounding, (c) transforming the rounded image by the DCT, (d) extracting the embedded watermark.

**TABLE 4.1**
**Total Possible Embedded Watermarks**

| Embedded | Extracted | Error Bits | Embedded | Extracted | Error Bits |
|---|---|---|---|---|---|
| 0000 | 0000 | 0 | 1000 | 0000 | 1 |
| 0001 | 0000 | 1 | 1001 | 0000 | 2 |
| 0010 | 0000 | 1 | 1010 | 0000 | 2 |
| 0011 | 0000 | 2 | 1011 | 0000 | 3 |
| 0100 | 0000 | 1 | 1100 | 0000 | 2 |
| 0101 | 0000 | 2 | 1101 | 0010 | 4 |
| 0110 | 0000 | 2 | 1110 | 1110 | 0 |
| 0111 | 1110 | 2 | 1111 | 1110 | 1 |

| – | – | C | D | – | – | – | – |
|---|---|---|---|---|---|---|---|
| – | B | – | – | – | – | – | – |
| A | – | – | – | – | – | – | – |
| – | – | – | – | – | – | – | – |
| – | – | – | – | – | – | – | – |
| – | – | – | – | – | – | – | – |
| – | – | – | – | – | – | – | – |
| – | – | – | – | – | – | – | – |

**FIGURE 4.13**   The positions where the watermark is embedded.

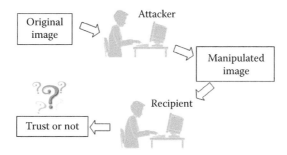

**FIGURE 4.14**   A recipient has a problem judging whether a received image is altered.

### 4.3.1   BLOCK-BASED FRAGILE WATERMARKS

Wong has presented a block-based fragile watermarking algorithm that is capable of detecting changes [14]—for example, to pixel values and image size—by adopting the Rivest–Shamir–Adleman (RSA) public-key encryption algorithm and message digest algorithm 5 (MD5) for the hash function [15,16]. Wong's watermarking insertion and extraction algorithms are introduced in the following, and their flowcharts are shown in Figures 4.15 and 4.16, respectively.

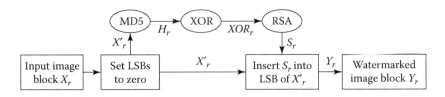

**FIGURE 4.15**　Wong's watermarking insertion algorithm.

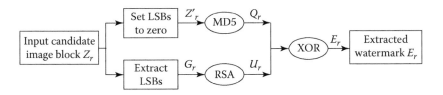

**FIGURE 4.16**　Wong's watermarking extraction algorithm.

*Wong's watermarking insertion algorithm:*

1. Divide the original image $X$ into subimages $X_r$.
2. Divide the watermark $W$ into subwatermarks $W_r$.
3. For each subimage $X_r$, we obtain $X_r'$ by setting the LSB to be 0.
4. For each $X_r'$, we obtain the corresponding codes $H_r$ with a cryptographic hash function (e.g., MD5).
5. $XOR_r$ is obtained by adopting the *exclusive or* (XOR) operation of $H_r$ and $W_r$.
6. $S_r$ is obtained by encrypting $XOR_r$ using the RSA encryption with a private key $K'$.
7. We obtain the watermarked subimage $Y_r$ by embedding $S_r$ into the LSB of $X_r'$.

*Wong's watermarking extraction algorithm:*

1. Divide the candidate image $Z$ into subimages $Z_r$.
2. $Z_r'$ is obtained by setting the LSB of $Z_r$ to be 0.
3. $G_r$ is obtained by extracting the LSB of $Z_r$.
4. $U_r$ is obtained by decrypting $G_r$ using RSA with a public key $K$.
5. For each $Z_r'$, we obtain the corresponding codes $Q_r$ with a cryptographic hash function (e.g., MD5 or SHA [secure hash algorithm]).
6. We obtain the extracted watermark $E_r$ by adopting the XOR operation of $Q_r$ and $G_r$.

### 4.3.2 WEAKNESS OF BLOCK-BASED FRAGILE WATERMARKS

Holliman and Memon have developed a VQ counterfeiting attack to forge the fragile watermark by exploiting the blockwise independence of Wong's algorithm [17]. Through this exploitation we can rearrange the blocks of an image to form a new collage image in which the embedded fragile watermark is not altered. Therefore, given a large database of watermarked images, the VQ attack can approximate a counterfeit collage image with the same visual appearance as an original unwatermarked image from a codebook. The attacker does not need to have knowledge of the embedded watermarks.

If $D$ is the large database of watermarked images and $C$ is the codebook with $n$ items generated from $D$, the algorithm of the VQ attack can be described as follows (its flowchart is shown in Figure 4.17):

*VQ counterfeiting attack algorithm*:

1. Generate the codebook $C$ from $D$.
2. Divide the original image $X$ into subimages $X_r$.
3. For each subimage $X_r$, we obtain $X_r^c$ by selecting the data from the codebook with the minimum distance (difference) to $X_r$. Finally, we obtain a counterfeit collage image $X^c$ with the same visual appearance as the original unwatermarked image.

### 4.3.3 HIERARCHY-BASED FRAGILE WATERMARKS

In order to defeat a VQ counterfeiting attack, a hierarchical block-based watermarking technique was developed by Celik et al. [18]. The idea of the hierarchical

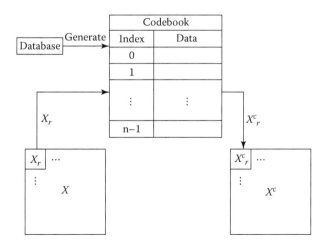

**FIGURE 4.17** The VQ counterfeit attack.

watermarking approach is simply to break the blockwise independence of Wong's algorithm. That is, the embedded data not only include the information of the corresponding block but also possess the relative information of the higher-level blocks. Therefore, a VQ attack cannot approximate an image based on a codebook that only records the information of each watermarked block.

Let $X_{i,j}^l$ denote a block in the hierarchical approach, where $(i,j)$ represents the spatial position of the block and $l$ is the level to which the block belongs. The total number of levels in the hierarchical approach is denoted by $L$. At each successive level, the higher-level block is divided into $2 \times 2$ lower-level blocks. That is, for $l = L-1$ to 2,

$$\begin{bmatrix} X_{2i,\ 2j}^{l+1} & X_{2i,\ 2j+1}^{l+1} \\ X_{2i+1,\ 2j}^{l+1} & X_{2i+1,\ 2j+1}^{l+1} \end{bmatrix} = X_{i,j}^l.$$

For each block $X_{i,j}^l$, we obtain its corresponding *ready-to-insert data* (RID), $S_{i,j}^l$, after the processes of MD5, XOR, and RSA. Then we construct a payload block $P_{i,j}^L$ based on the lowest-level block, $X_{i,j}^L$. For each payload block, it contains both the RID and the data belonging to the higher-level block of $X_{i,j}^L$.

Figure 4.18 illustrates an example of the hierarchical approach; Figure 4.18a denotes an original image $X$ and its three-level hierarchical results. Note that $X_{0,0}^1$ is the top level of the hierarchy, consisting of only one block, $X$. Figure 4.18b shows the corresponding RID of each block, $X_{i,j}^l$. In Figure 4.18c, when dealing with the block

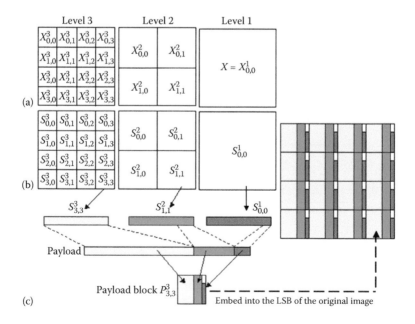

**FIGURE 4.18**   An example of the hierarchical approach with three levels.

$X_{3,3}^3$, we consider the following three RIDs: $S_{3,3}^3$, $S_{1,1}^2$, and $S_{0,0}^1$. After generating the payload block $P_{3,3}^3$, we embed it into the LSB of the original image $X$ on $X_{3,3}^3$.

## 4.4 ROBUST WATERMARKS

Computer technology allows people to easily duplicate and distribute digital multimedia through the Internet. However, these benefits come with the concomitant risks of data piracy. One means of providing security for copyright protection is robust watermarking, which ensures that the embedded messages survive attacks such as JPEG compression, Gaussian noise, and low-pass filtering. Figure 4.19 illustrates the purpose of robust watermarking.

### 4.4.1 REDUNDANT EMBEDDING APPROACH

It is obvious that the embedded message will be distorted due to the image-processing procedures of attackers. In order to achieve robustness, we embed redundant watermarks in a host image [19]. Epstein and McDermott hold a patent for copy protection via redundant watermark encoding [20].

### 4.4.2 SPREAD SPECTRUM

Cox et al. have presented a spread spectrum–based robust watermarking algorithm in which the embedded messages are spread throughout the image [21,22]. Their algorithm first translates an image into its frequency domain; then the watermarks are embedded into the significant coefficients of the transformed image. The significant coefficients are the locations of the large absolute values of the transformed image, as shown in Figure 4.20. Figure 4.20a shows an original image, and Figure 4.20b is the image transformed by the DCT. The threshold is 70. The bold rectangles in Figure 4.20b are called the *significant coefficients* since the absolute values are larger than 70.

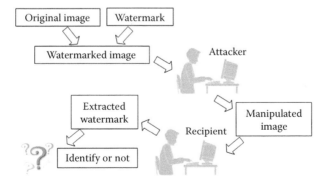

**FIGURE 4.19**   The purpose of robust watermarking.

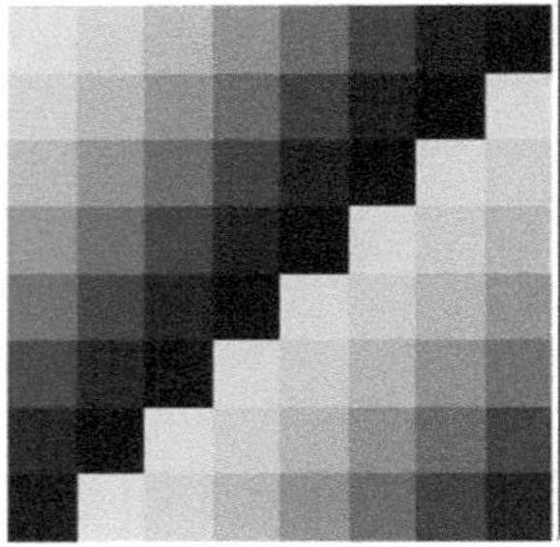

| 215 | 201 | 177 | 145 | 111 | 79 | 55 | 41 |
|-----|-----|-----|-----|-----|-----|-----|-----|
| 201 | 177 | 145 | 111 | 79 | 55 | 41 | 215 |
| 177 | 145 | 111 | 79 | 55 | 41 | 215 | 201 |
| 145 | 111 | 79 | 55 | 41 | 215 | 201 | 177 |
| 111 | 79 | 55 | 41 | 215 | 201 | 177 | 145 |
| 79 | 55 | 41 | 215 | 201 | 177 | 145 | 111 |
| 55 | 41 | 215 | 201 | 177 | 145 | 111 | 79 |
| 41 | 215 | 201 | 177 | 145 | 111 | 79 | 55 |

| 1024.0 | 0.0 | 0.0 | 0.0 | 0.0 | 0.0 | 0.0 | 0.0 |
|--------|-----|-----|-----|-----|-----|-----|-----|
| 0.0 | 110.8 | 244.0 | -10.0 | 35.3 | -2.6 | 10.5 | -0.5 |
| 0.0 | 244.0 | -116.7 | -138.2 | 0.0 | -27.5 | 0.0 | -6.4 |
| 0.0 | -10.0 | -138.2 | 99.0 | 72.2 | 0.2 | 13.3 | 0.3 |
| 0.0 | 35.3 | 0.0 | 72.2 | -94.0 | -48.2 | 0.0 | -7.0 |
| 0.0 | -2.6 | -27.5 | 0.2 | -48.2 | 92.3 | 29.9 | 0.4 |
| 0.0 | 10.5 | 0.0 | 13.3 | 0.0 | 29.9 | -91.3 | -15.7 |
| 0.0 | -0.5 | -6.4 | 0.3 | -7.0 | 0.4 | -15.7 | 91.9 |

**FIGURE 4.20** An example of significant coefficients.

## REFERENCES

1. Podilchuk, C. I. and Delp, E. J., Digital watermarking: Algorithms and applications, *IEEE Signal Processing Mag.*, 18, 33, 2001.
2. Pitas, I., A method for watermark casting on digital images, *IEEE Trans. Circuits and Systems for Video Technology*, 8, 775, 1998.
3. Wolfgang, R. and Delp, E., A watermarking technique for digital imagery: Further studies, in *Int. Conf. Imaging Science, Systems and Technology*, Las Vegas, NV, 1997, 279.
4. Shih, F. Y. and Wu, Y., Combinational image watermarking in the spatial and frequency domains, *Pattern Recognition*, 36, 969, 2003.
5. Wong, P., A watermark for image integrity and ownership verification, in *Proc. Int. Conf. IS&T PICS*, Portland, OR, 1998, 374.
6. Wolfgang, R. and Delp, E., A Watermark for digital images, in *Proc. IEEE Int. Conf. Image Processing*, Lausanne, Switzerland, 1996, 219.
7. Lin, E. and Delp, E., Spatial synchronization using watermark key structure, in *Proc. SPIE Conf. Security, Steganography, and Watermarking of Multimedia Contents*, San Jose, CA, 2004, 536.
8. Mukherjee, D., Maitra, S., and Acton, S., Spatial domain digital watermarking of multimedia objects for buyer authentication, *IEEE Trans. Multimedia*, 6, 1, 2004.
9. Lin, S. and Chen, C., A robust DCT-based watermarking for copyright protection, *IEEE Trans. Consumer Electronics*, 46, 415, 2000.
10. Huang, J., Shi, Y., and Shi, Y., Embedding image watermarks in DC components, *IEEE Trans. Circuits and Systems for Video Technology*, 10, 974, 2000.
11. Cox, J. et al., A secure, robust watermark for multimedia, in *Proc. First Int. Workshop Information Hiding*, Cambridge, UK, 1996, 185.
12. Gray, R. M., Vector quantization, *IEEE ASSP Mag.*, 1, 4, 1984.
13. Lu, Z. and Sun, S., Digital image watermarking technique based on vector quantization, *IEEE Electronics Letters*, 36, 303, 2000.
14. Wong, P. W., A public key watermark for image verification and authentication, in *Proc. IEEE Int. Conf. Image Processing*, Chicago, IL, 1998, 425.
15. Rivest, R., Shamir, A., and Adleman, L., A method for obtaining digital signatures and public-key cryptosystems, *Communications of the ACM*, 21, 120, 1978.
16. Rivest, R. L., The MD5 message digest algorithm, in *RFC 1321*, MIT Laboratory for Computer Science and RSA Data Security, 1992.
17. Holliman, M. and Memon, N., Counterfeiting attacks on oblivious block-wise independent invisible watermarking schemes, *IEEE Trans. Image Processing*, 9, 432, 2000.
18. Celik, M. et al., Hierarchical watermarking for secure image authentication with localization, *IEEE Trans. Image Processing*, 11, 585, 2002.
19. Coetzee, L. and Eksteen, J., Copyright protection for cultureware preservation in digital repositories, in *Proc. 10th Annual Internet Society Conf.*, Yokohama, Japan, 2000.
20. Epstein, M. and McDermott, R., Copy protection via redundant watermark encoding, USA Patent no. 7133534, 2006, accessed on January 5, 2017, http://www.patentgenius.com/patent/7133534.html.
21. Cox, I. J. et al., Secure spread spectrum watermarking for multimedia, *IEEE Trans. Image Processing*, 6, 1673, 1997.
22. Cox, I. J. et al., Digital watermarking, *J. Electronic Imaging*, 11, 414, 2002.

# 5 Watermarking Attacks and Tools

There are a great number of digital watermarking methods developed in recent years that embed secret watermarks in images for various applications. The purpose of watermarking can be *robust* or *fragile*, *private* or *public*. A robust watermark aims to overcome the attacks from common image processing or intentional attempts to remove or alter it [1–3]. A fragile watermark aims to sense whether the watermarked image has been altered—even slightly. A private watermark is used when the original unmarked image is available for comparison. However, a public watermark is set up in an environment in which the original unmarked image is unavailable for watermark detection.

Attackers of watermarked images can be classified as unintentional or intentional. An attack action will disable a user's watermarking target in a way that the embedded message cannot function well. An unintentional attacker may not be aware of affecting the original watermark; for example, one could reduce an image's size through JPEG compression or may choose to smooth out the image a bit. An intentional attacker malignantly disables the function of watermarks for a specific reason; for example, one could alter or destroy the hidden message for an illegal purpose such as image forgery.

In order to identify the weaknesses of watermarking techniques, propose improvements, and study the effects of current technology on watermarks, we need to understand the varieties of attacks on watermarked images. In this chapter, we will discuss different types of watermarking attacks and available tools. Since there are many different types of attacks for a variety of purposes, not everything will be covered here, but examples will be given of some popular attacking schemes. We can categorize these various watermarking attack schemes into four classes: image-processing attacks, geometric transformation, cryptographic attacks, and protocol attacks.

This chapter is organized as follows. Sections 5.1–5.4 introduce image processing, geometric transformation, cryptographic attacks, and protocol attacks, respectively. Section 5.5 describes watermarking tools. Section 5.6 presents an efficient block-based fragile watermarking system for tamper localization and recovery.

## 5.1 IMAGE-PROCESSING ATTACKS

Image-processing attacks include filtering, remodulation, JPEG coding distortion, and JPEG 2000 compression.

### 5.1.1 ATTACK BY FILTERING

Filtering indicates an image-processing operation in the frequency domain. We first compute the Fourier transform of the image, multiply the result by a filter transfer function, and then perform the inverse Fourier transform. The idea of sharpening is to increase the magnitude of high-frequency components relative to low-frequency components. A high-pass filter in the frequency domain is equivalent to an impulse shape in the spatial domain. Therefore, a sharpening filter in the spatial domain contains positive values close to the center and negative values surrounding the outer boundary. A typical $3 \times 3$ sharpening filter is

$$\frac{1}{9}\begin{bmatrix} -1 & -1 & -1 \\ -1 & 8 & -1 \\ -1 & -1 & -1 \end{bmatrix}. \tag{5.1}$$

An example of a $3 \times 3$ sharpening filter attack is shown in Figure 5.1.

The idea of blurring is to decrease the magnitude of high-frequency components. A low-pass filter in the frequency domain is equivalent to a mountain shape in the spatial domain. Therefore, a smoothing filter in the spatial domain should have all positive coefficients, with the largest in the center. The simplest low-pass filter would be a mask with all coefficients having a value of 1. A sample $3 \times 3$ Gaussian filter is

$$\frac{1}{12}\begin{bmatrix} 1 & 1 & 1 \\ 1 & 4 & 1 \\ 1 & 1 & 1 \end{bmatrix}. \tag{5.2}$$

A sample $5 \times 5$ Gaussian filter is

$$\frac{1}{273}\begin{bmatrix} 1 & 4 & 7 & 4 & 1 \\ 4 & 16 & 26 & 16 & 4 \\ 7 & 26 & 41 & 26 & 7 \\ 4 & 16 & 26 & 16 & 4 \\ 1 & 4 & 7 & 4 & 1 \end{bmatrix}. \tag{5.3}$$

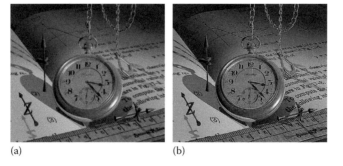

(a)                              (b)

**FIGURE 5.1**    (a) The watermarked image, (b) after the $3 \times 3$ sharpening filter attack.

(a)                              (b)

**FIGURE 5.2** (a) The watermarked image, (b) after the $5 \times 5$ Gaussian filter attack.

(a)                              (b)

**FIGURE 5.3** (a) The watermarked image, (b) after the $5 \times 5$ median filtering attack.

An example of a $5 \times 5$ Gaussian filter attack is shown in Figure 5.2.

In image enhancement, a median filter is often used to achieve noise reduction rather than blurring. The method is rather simple: we calculate the median of the gray values surrounding the pixel's neighborhood and assign it to the pixel. An example of a $5 \times 5$ median filter is shown in Figure 5.3.

### 5.1.2 ATTACK BY REMODULATION

An ancient remodulation attack was first presented by Langelaar et al. [4]. In this scheme the watermark was predicted via subtraction of the median-filtered version of the watermarked image. The predicted watermark was additionally high-pass filtered, truncated, and then subtracted from the watermarked image with a constant amplification factor of 2. The basic remodulation attack consists of denoising the image, inverting the estimated watermark, and adding it to a percentage of certain locations. If the watermark is additive, this attack is quite effective and leaves few artifacts. In most cases the attacked image contains little distortion. An example of a remodulation attack is shown in Figure 5.4.

(a)                                    (b)

**FIGURE 5.4**    (a) The watermarked image, (b) after the $3 \times 3$ remodulation attack. (From Pereira, S. et al., *Proc. Information Hiding Workshop III*, Pittsburgh, PA, 2001.)

### 5.1.3    ATTACK BY JPEG CODING DISTORTION

JPEG, short for the Joint Photographic Experts Group, is very popular in color image compression. It was created in 1986, issued in 1992, and approved by the ISO (International Organization for Standardization) in 1994. Often used in Internet browsers is the progressive JPEG, which reorders the information in such a way that, after only a small part of the image has been downloaded, an obscure perception of the whole image is available rather than a brittle perception of just a small portion. Although JPEG can reduce an image file to about 5% of its normal size, some detail may be lost in the compressed image. Figure 5.5 shows a JPEG-encoded version of *Lena* with a quality of 10.

### 5.1.4    ATTACK BY JPEG 2000 COMPRESSION

JPEG 2000, which is a standard image compression technique created by the JPEG committee, uses the state-of-the-art wavelet-based technology. This creation aims at overcoming the blocky appearance problem occurred in the discrete cosine transform–based JPEG standard. Its architecture has various applications ranging from prepublishing, digital cameras, medical imaging, and other industrial imaging. In

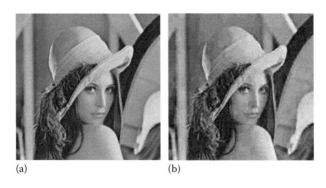

(a)                                    (b)

**FIGURE 5.5**    (a) The watermarked image, (b) after the JPEG attack.

(a)                                              (b)

**FIGURE 5.6**    (a) The watermarked image, (b) after the JPEG 2000 attack.

general, JPEG 2000 is able to achieve a high compression ratio and simultaneously avoid blocky and blurry artifacts. It also allows more sophisticated progressive downloads. An example of a JPEG 2000 attack with 0.5 bits per pixel (BPP) is shown in Figure 5.6.

## 5.2    GEOMETRIC ATTACKS

Geometric attacks include scaling, rotation, clipping, linear transformation, bending, warping, perspective projection, collage, and template. O'Ruanaidh and Pun proposed digital watermarking using Fourier–Mellin transform–based invariants [6]. They do not need the original image to extract the embedded watermark. According to their design, the watermark will not be destroyed by combined rotation, scale, and translation transformations. The original image is not required for extracting the embedded mark.

### 5.2.1    ATTACK BY IMAGE SCALING

Sometimes, when we scan an image hardcopy or when we adjust the size of an image for electronic publishing, image scaling may happen. This should be especially noticed as we move more and more toward Web publishing. An example is when the watermarked image is first scaled down (or down-sampled) by reducing both the length and width by half—that is, by averaging every $2 \times 2$ block into a pixel. Then, this quarter-sized image is scaled up (or up-sampled) to its original size—that is, by the interpolation method. In general, bilinear interpolation will yield a smooth appearance. Bicubic interpolation will yield even smoother results. An example of a scaling attack is shown in Figure 5.7.

Image scaling can have two types: uniform and nonuniform scaling. Uniform scaling uses the same scaling factors in the horizontal and vertical directions. However, nonuniform scaling applies different scaling factors in the horizontal and vertical directions. In other words, the aspect ratio is changed in nonuniform scaling. An example of nonuniform scaling using $x$ scale $= 0.8$ and $y$ scale $= 1$ is shown in Figure 5.8. Note that nonuniform scaling will produce the effect of object shape distortion. Most digital watermarking methods are designed to be flexible only to uniform scaling.

(a)                              (b)

**FIGURE 5.7**   (a) The watermarked image, (b) after the scaling attack.

(a)                              (b)

**FIGURE 5.8**   (a) The watermarked image, (b) after the nonuniform scaling attack. (From Pereira, S. et al., *Proc. Information Hiding Workshop III*, Pittsburgh, PA, 2001.)

## 5.2.2   ATTACK BY ROTATION

Suppose that we want to apply a rotation angle $\theta$ to an image $f(x,y)$. Let the counterclockwise rotation angle be positive and the image after rotation be $f^*(x^*,y^*)$. Figure 5.9 illustrates the rotation from a point $P$ to $P^*$. Let $R$ denote the distance from $P$ to the origin. We can derive the following formulae:

$$x = r\cos\varphi, \quad y = r\sin\varphi. \tag{5.4}$$

$$x^* = r\cos(\theta + \varphi) = r\cos\varphi\cos\theta - r\sin\varphi\sin\theta$$
$$= x\cos\theta - y\sin\theta. \tag{5.5}$$

$$y^* = r\sin(\theta + \varphi) = r\cos\varphi\sin\theta + r\sin\varphi\cos\theta$$
$$= x\sin\theta + y\cos\theta. \tag{5.6}$$

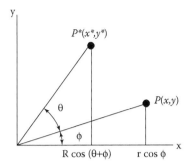

**FIGURE 5.9**   Illustration of the rotation.

$$\begin{bmatrix} x^* & y^* \end{bmatrix} = \begin{bmatrix} x & y \end{bmatrix} \begin{bmatrix} \cos\theta & \sin\theta \\ -\sin\theta & \cos\theta \end{bmatrix}. \tag{5.7}$$

When we perform the formulae to compute the new coordinates, they need to be converted into integers. A simple rounding approach will often produce unacceptable results, since some points of the new image will have no gray levels assigned to them by this method, while others will have more than one gray level assigned. To solve these problems, the inverse rotation and interpolation methods are needed [7].

When we apply a small angle rotation, often combined with cropping, to a watermarked image, the overall perceived content of the image is the same, but the embedded watermark may be undetectable. When an image hardcopy is scanned, a rotation is often required to adjust a slight tilt. An example of applying a 10° counterclockwise rotation to a watermarked image is shown in Figure 5.10.

(a)                              (b)

**FIGURE 5.10**   (a) The watermarked image, (b) after the rotational attack. (From Pereira, S. et al., *Proc. Information Hiding Workshop III*, Pittsburgh, PA, 2001.)

### 5.2.3 ATTACK BY IMAGE CLIPPING

Figure 5.11a shows a clipped version of the watermarked image. Only the central quarter of the image remains. In order to extract the watermark from this image, the missing portions of the image are replaced by the corresponding portions from the original unwatermarked image, as shown in Figure 5.11b.

### 5.2.4 ATTACK BY LINEAR TRANSFORMATION

We can apply a $2 \times 2$ linear transformation matrix. Let $x$ and $y$ denote the original coordinates, and $x^*$ and $y^*$ denote the transformed coordinates. The transformation matrix $T$ to change $(x,y)$ into $(x^*,y^*)$ can be expressed as

$$\begin{bmatrix} x^* & y^* \end{bmatrix} = \begin{bmatrix} x & y \end{bmatrix} \cdot \begin{bmatrix} T_{11} & T_{12} \\ T_{21} & T_{22} \end{bmatrix}. \tag{5.8}$$

An example of using $T_{11} = 1.15$, $T_{12} = -0.02$, $T_{21} = -0.03$, and $T_{22} = 0.9$ is shown in Figure 5.12.

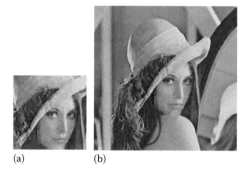

(a)　　　　(b)

**FIGURE 5.11**　(a) Clipped version of the JPEG-encoded (10% quality) *Lena*, (b) the restored version.

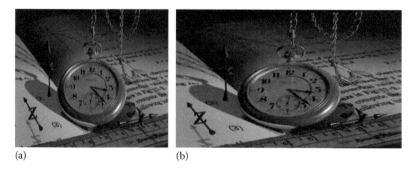

(a)　　　　　　　(b)

**FIGURE 5.12**　(a) The watermarked image, (b) after the linear transformation attack. (From Pereira, S. et al., *Proc. Information Hiding Workshop III*, Pittsburgh, PA, 2001.)

## 5.2.5 ATTACK BY BENDING

This attack was originally proposed by Petitcolas [8]. Local nonlinear geometric distortions are applied and the image is interpolated. Noise is also added and a small compression is applied. An example of a bending attack is shown in Figure 5.13.

## 5.2.6 ATTACK BY WARPING

Image warping performs a pixel-by-pixel remapping (or *warp*) of an input image to an output image. It requires an integer-sampled *remapping image* that is the same size as the output image. This remapping image consists of *x* and *y* channels. For each pixel in the output image, the corresponding point in the remapping image is checked to determine which pixel from the input image it will be copied to. The image grid is first warped into three dimensions and then projected back onto two dimensions. An example of a warping attack with a factor of 3 is shown in Figure 5.14.

## 5.2.7 ATTACK BY PERSPECTIVE PROJECTION

In contrast to parallel projection, in perspective projection the parallel lines converge, the object size is reduced with increasing distance from the center of projection, and the nonuniform foreshortening of lines in the object occurs as a function of

(a)      (b)

**FIGURE 5.13** (a) The watermarked image, (b) after the bending attack. (From Pereira, S. et al., *Proc. Information Hiding Workshop III*, Pittsburgh, PA, 2001.)

(a)      (b)

**FIGURE 5.14** (a) The watermarked image, (b) after the warping attack. (From Pereira, S. et al., *Proc. Information Hiding Workshop III*, Pittsburgh, PA, 2001.)

orientation and the distance of the object from the center of projection. All of these effects aid the depth perception of the human visual system, but the object shape is not preserved. An example of an attack using 30° rotation along the *x*-axis together with a perspective projection is shown in Figure 5.15.

### 5.2.8 ATTACK BY COLLAGE

The collage attack was proposed by Holliman and Memon [9] on blockwise independent watermarking techniques. They assume that the watermark logo is available to the attacker. The multiple authenticated images are used to develop a new image. They combine parts of different images with a preservation of their relative spatial locations in the image. Each block is embedded independently. An example of a collage attack is shown in Figure 5.16.

### 5.2.9 ATTACK BY TEMPLATES

Templates are used as a synchronization pattern to estimate the applied affine transformation [10]. Then we apply the inverse transformation to extract the watermark

(a)                              (b)

**FIGURE 5.15**   (a) The watermarked image, (b) after the perspective projection attack. (From Pereira, S. et al., *Proc. Information Hiding Workshop III*, Pittsburgh, PA, 2001.)

(a)                              (b)

**FIGURE 5.16**   (a) The watermarked image, (b) after the collage attack. (From Pereira, S. et al., *Proc. Information Hiding Workshop III*, Pittsburgh, PA, 2001).

based on the obtained estimation. Many modern schemes use a template in the frequency domain to extract the watermark. The template is composed of peaks. The watermark detection process consists of two steps. First, the affine transformation undergone by the watermarked image is calculated. Second, the inverse transformation is applied to decode the watermark.

## 5.3   CRYPTOGRAPHIC ATTACKS

The principles of cryptographic attacks are similar to those applied in cryptography but with different purposes. A cryptographic attack attempts to circumvent the security of a cryptographic system by retrieving the key or exposing the plaintext in a code, cipher, or protocol. Based on exhaustive key search approaches, the cryptographic attack method tries to find out the key used for the embedding. When the key is found, the watermark is overwritten. Another cryptographic attack called an *oracle attack* tries to create a nonwatermarked image when a watermark detector device is available. The attacked data set is generated by attacking a small portion of each data set and establishing a new data set for attacking. In other words, an attacker can remove a watermark if a public decoder is available by applying small changes to the image until the decoder cannot find it anymore. Linnartz and Dijk [11] presented an analysis of a sensitivity attack against electronic watermarks in images.

Another cryptographic attack called *collusion* is used when instances of the same data are available. An example of attack by collusion is to take two separately watermarked images and average them to form Figure 5.17.

## 5.4   PROTOCOL ATTACK

Protocol attacks intend to generate protocol ambiguity in the watermarking process. An example is using invertible watermarks to subtract a designed watermark from an image and then declaring it the owner. This can cause ambiguity regarding true ownership. One protocol attack, called a *copy attack* [12], intends to estimate a watermark from watermarked data and copy it to some other data, instead of destroying the watermark or impairing its detection. A copy attack consists of four stages: (1)

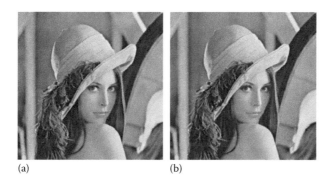

(a)                                              (b)

**FIGURE 5.17**    (a) The watermarked image, (b) after the collusion attack.

an estimation of the watermark from the original image, (2) an estimation of the mask from the target image, (3) the reweighting of the watermark as a function of the mask, and (4) the addition of the reweighted watermark onto the target image.

## 5.5  WATERMARKING TOOLS

This section briefly describes four benchmarking tools for image watermarking: Stirmark, Checkmark, Optimark, and Certimark.

Stirmark was developed in November 1997 as a generic tool for evaluating the robustness of image-watermarking algorithms. It applied random bilinear geometric distortions in order to desynchronize watermarking algorithms. Since then, enhanced versions have been developed to improve the original attack. Stirmark can generate a set of altered images based on the watermarked image to identify whether the embedded watermark is detectable [8,13–15]. Stirmark uses a combination of different detection results to determine an overall score. There are nine types of attacks available in Stirmark: compression, cropping, scaling, shearing, rotation, signal enhancement, linear transformation, random geometric distortion, and other geometric transformations.

Checkmark [5], which can run on MATLAB® under Unix and Windows, provides a benchmarking tool to evaluate the performance of image-watermarking techniques. It includes some additional attacks that are unavailable in Stirmark. It has the characteristic of considering different applications in image watermarking and placing different weights based on their importance for a particular usage.

Optimark [16], which was developed at Aristotle University in Thessaloniki, Greece, can evaluate the performance of still image–watermarking algorithms. Its main features include multiple trials of different messages and keys, evaluations of detection/decoding metrics, a graphical user interface, an algorithm breakdown evaluation, and a resulting summarization of selected attacks.

Certimark was developed by a European project including 15 academic and industrial partners in 2000. It intends to perform an international certification process on watermarking algorithms as a benchmark reference, and is used as a basis for accessing the control of a given watermarking technique. It is believed that the clear framework in the technological assessment will lead to competition between technology suppliers and in the meantime will warrant the robustness of watermarking as measured by the benchmark.

## 5.6  AN EFFICIENT BLOCK-BASED FRAGILE WATERMARKING SYSTEM FOR TAMPER LOCALIZATION AND RECOVERY

In this section we present an efficient block-based fragile watermarking system for tamper localization and the recovery of images. We apply the cyclic redundancy checksum (CRC) to authenticate the features of a block stored in a pair of mapping blocks. The proposed scheme localizes the tampered area irrespective of the gray value, and the detection strategy is designed in such a way as to reduce false alarms. It can be applied to grayscale and color images. We conduct malicious attacks using different cropping patterns, sizes, and shapes. Experimental results show that the proposed system outperforms existing schemes in terms of the recovery rate of modified blocks, support against vector quantization attacks, the localization of the tampered area, and the reduction of false alarms.

Section 5.6.1 presents work related to our method. Our proposed method is presented in Section 5.6.2. Experimental results and discussions are presented in Section 5.6.3. Finally, conclusions are drawn in Section 5.6.4.

### 5.6.1 RELATED WORK

#### 5.6.1.1 Vector Quantization Attack

Holliman and Memon [9] presented a counterfeiting attack on blockwise independent watermarking schemes. In this scheme the attacker generates a forged image by the collaging of authenticated blocks from different watermarked images. Because of the blockwise nature of the embedding and authentication processes, the forgery is authenticated as valid. Additionally, if the database of watermarked images is huge, the attacker can easily generate a collage image that is identical to the unwatermarked image.

#### 5.6.1.2 Cyclic Redundancy Checksum

In the computation of CRC [17,18], a $k$-bit string is represented by the coefficients of a $k-1$-degree polynomial—that is, the coefficients of $x^0$ to $x^{k-1}$. The most significant bit (MSB) of the $k$-bit string becomes the coefficient of $x^{k-1}$, the second MSB becomes the coefficient of $x^{k-2}$, and so on, and the least-significant bit (LSB) becomes the coefficient of $x^0$. For example, the polynomial representation of the bit string 1011 is $x^3 + x + 1$. To compute the CRC-$r$ authentication information of a bit string, we pad $r$ zeros onto the end of the bit string. Then we divide the resulting bit string with an $r$-degree polynomial. Then we subtract the remainder from the bit string padded with zeros. The resultant bit string after subtraction is the authentication information. The subtraction is done in modulo-2 arithmetic. To check if a CRC-$r$-generated bit string is modified or not, we divide the bit string with the $r$-degree polynomial used in the generation stage. If there is a remainder after division, it means that the bit string is modified; otherwise, the bit string is not manipulated.

#### 5.6.1.3 Comparison of Existing Methodologies

In Table 5.1 a comparison of three existing blockwise-dependent fragile watermarking schemes is made using different characteristics. The three schemes considered are Lin et al. [17], Lin et al. [19], and Lee and Lin [20]. Our proposed algorithm is designed to achieve the best characteristics of these three schemes—that is, a tamper localization size of $2 \times 2$, two copies of the recovery information, support against VQ attacks, and the detection of modification in higher-order bits. Our proposed scheme can be applied to grayscale and color images. In the proposed algorithm the probability of thwarting VQ attacks and higher-order bit modification is higher than that of Lin et al. [17] and Lin et al. [19] because of the two copies of recovery information. In order to achieve dual recovery we adopt the generation of a lookup table, similar to [20].

**TABLE 5.1**

**Comparison of Three Existing Schemes**

| Characteristic | Lin et al. [17] | Lin et al. [19] | Lee and Lin [20] |
|---|---|---|---|
| Minimum block size of tamper localization | 8 × 8 | 4 × 4 | 2 × 2 |
| Copies of recovery information | 1 | 1 | 2 |
| Detection of VQ attack | Yes | Yes | No |
| Number of bits modified in each pixel of the original image | 2 (LSB) | 2 (LSB) | 3 (LSB) |
| Blockwise-dependent detection | Yes | Yes | No |
| Detection of changes in the higher-order bits of each pixel | Yes, if the corresponding mapping block is valid | Yes, if the corresponding mapping block is valid | No |
| Authentication method used | MD5 with CRC | Parity bit | Parity bit |

### 5.6.1.4  Lookup Table Generation

We outline the procedure of the lookup table generation as follows:

1. Let $A$ be a host image of size $M \times M$, where $M$ is a multiple of 2.
2. Divide the image $A$ into blocks of size $2 \times 2$. Calculate the block number for each block as

$$X(i,j) = (i-1) \times (M/2) + (j-1), \tag{5.9}$$

where $i, j \in [1, M/2 \times M/2]$.

3. Calculate the corresponding mapping block for each block number as

$$X'(i,j) = \{(k \times X(i,j)) \bmod N\} + 1, \tag{5.10}$$

where:
$i, j \in [1, M/2 \times M/2]$; $X, X' \in [0, N-1]$ are the corresponding one-to-one mapping block numbers
$k \in [2, N-1]$ (secret key) is a prime number
$N = M/2 \times M/2$ is the total number of blocks in the image
Note that if $k$ is not a prime number, we will not achieve one-to-one mapping.

4. Shift the columns in $X'$, so the blocks that belong to the left half of the image are moved to the right and the blocks that belong to the right half of the image are moved to the left. Denote the new lookup table as $X''$. Figure 5.18 shows the generation of a lookup table for a $16 \times 16$ image with $k = 3$.

| 0 | 1 | 2 | 3 | 4 | 5 | 6 | 7 |
|---|---|---|---|---|---|---|---|
| 8 | 9 | 10 | 11 | 12 | 13 | 14 | 15 |
| 16 | 17 | 18 | 19 | 20 | 21 | 22 | 23 |
| 24 | 25 | 26 | 27 | 28 | 29 | 30 | 31 |
| 32 | 33 | 34 | 35 | 36 | 37 | 38 | 39 |

(a)

| 1 | 4 | 7 | 10 | 13 | 16 | 19 | 22 |
|---|---|---|---|---|---|---|---|
| 25 | 28 | 31 | 34 | 37 | 40 | 43 | 46 |
| 49 | 52 | 55 | 58 | 61 | 0 | 3 | 6 |
| 9 | 12 | 15 | 18 | 21 | 24 | 27 | 30 |
| 33 | 36 | 39 | 42 | 45 | 48 | 51 | 54 |

(b)

| 4 | 7 | 13 | 22 | 1 | 10 | 16 | 19 |
|---|---|---|---|---|---|---|---|
| 28 | 31 | 37 | 46 | 25 | 34 | 40 | 43 |
| 52 | 55 | 61 | 6 | 49 | 58 | 0 | 3 |
| 12 | 15 | 21 | 30 | 9 | 18 | 24 | 27 |
| 36 | 39 | 45 | 54 | 33 | 42 | 48 | 51 |

(c)

**FIGURE 5.18** (a) Original table, (b) the mapping table obtained by Equation 5.10, (c) the lookup table after column transformation.

## 5.6.2 PROPOSED METHOD

The design of our proposed scheme is divided into three modules as watermark embedding, tamper detection with localization, and the recovery of tampered blocks.

### 5.6.2.1 Watermark Embedding

The watermark-embedding process consists of the following six steps:

1. Let $A$ be a host image of size $M \times M$, where $M$ is a multiple of 2. Compute the block numbers ($X$) and lookup table ($X''$), as described in Section 5.6.1.4.
2. Divide the host image $A$ horizontally into two halves, as shown in Figure 5.19. The corresponding blocks in the two halves form a partner block—that is, blocks 0&32, 1&33, 16&48, and so on.
3. For each partner block $a$-$b$ in the host image $A$, we set the three LSBs of each pixel in the blocks $a$ and $b$ to zero. Compute the average intensities $avg\_a$ and $avg\_b$, as shown in Figure 5.20.
4. Determine $x$-$y$, the corresponding partner block to $a$-$b$, from the lookup table $X$. The 10-bit feature of partner block $a$-$b$ is shown in Figure 5.21. Add two zeros to the end of the feature computed. We then use the CRC to compute authentication information. Obtain CRC-2 by dividing the 12-bit feature with a secret polynomial of degree 2 and subtracting the remainder from the feature.

| 0 | 1 | 2 | 3 | 4 | 5 | 6 | 7 |
|---|---|---|---|---|---|---|---|
| 8 | 9 | 10 | 11 | 12 | 13 | 14 | 15 |
| 16 | 17 | 18 | 19 | 20 | 21 | 22 | 23 |

| 4 | 7 | 13 | 22 | 1 | 10 | 16 | 19 |
|---|---|---|---|---|---|---|---|
| 28 | 31 | 37 | 46 | 25 | 34 | 40 | 43 |
| 52 | 55 | 61 | 6 | 49 | 58 | 0 | 3 |

| 32 | 33 | 34 | 35 | 36 | 37 | 38 | 39 |
|---|---|---|---|---|---|---|---|
| 40 | 41 | 42 | 43 | 44 | 45 | 46 | 47 |
| 48 | 49 | 50 | 51 | 52 | 53 | 54 | 55 |

| 36 | 39 | 45 | 54 | 33 | 42 | 48 | 51 |
|---|---|---|---|---|---|---|---|
| 60 | 63 | 5 | 14 | 57 | 2 | 8 | 11 |
| 20 | 23 | 29 | 38 | 17 | 26 | 32 | 35 |

(a)                                          (b)

**FIGURE 5.19**    (a) Two-halves of a $16 \times 16$ image with block numbers, (b) two-halves of the corresponding lookup table generated with $k = 3$.

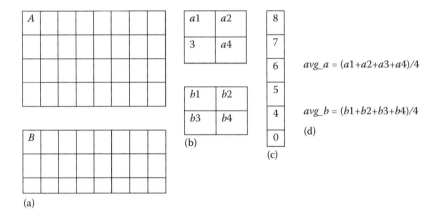

(a)                (b)                (c)                (d)

$avg\_a = (a1+a2+a3+a4)/4$

$avg\_b = (b1+b2+b3+b4)/4$

**FIGURE 5.20**    (a) The partner block $a\text{-}b$ of image $A$, (b) $2 \times 2$ view of $a\text{-}b$, (c) replacing three LSBs of each pixel with zero, (d) computation of the average intensity.

5. The 12-bit watermark of the partner block $a\text{-}b$ is shown in Figure 5.22b. Embed the computed 12-bit watermark into the three LSBs of each pixel of the corresponding mapping partner block $x\text{-}y$, as shown in Figure 5.22c. Note that the same watermark is embedded into two blocks, that is, $x$ and $y$.

6. After completing the embedding process for each partner block in the host image $A$, we will obtain the watermarked image $B$.

### 5.6.2.2   Tamper Detection

Tamper detection is done in four levels. In each level we try to reduce false alarms along with the localization of the tampered area. We denote the image under detection as $B'$.

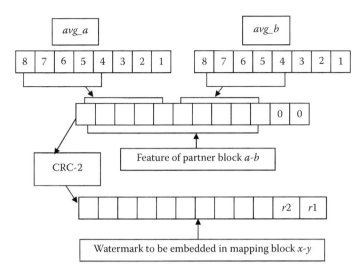

**FIGURE 5.21**   Feature of partner block *a-b*.

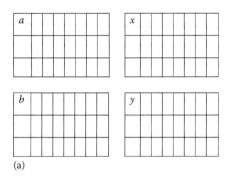

**FIGURE 5.22**   (a) Partner block *a-b* and its mapping partner block *x-y*, (b) 12-bit watermark, (c) embedding sequence of 12-bit watermark in each block of the mapping partner block.

1. The image under the detection process is divided into blocks of size $2 \times 2$, and the mapping sequence is generated as explained in Section 5.6.1.4.

2. *Level-1 detection*: For each block $b$, the 12-bit watermark is extracted from the three LSBs of each pixel. Divide the extracted watermark with the secret polynomial used in the embedding stage. If there is a remainder, it means that the block has been tampered with in the three LSBs, so mark $b$ as invalid; otherwise, mark $b$ as valid. Level 2 and Level 3 are used to reduce the false alarms.

3. *Level-2 detection*: For each valid block $b$ after Level-1 detection, we conduct the following:

> If $b$ is a block in the border of the image, then
>
>> if the number of invalid blocks in the $3 \times 3$ neighborhood of $b \geq 3$,
>>
>>> mark $b$ as invalid.
>
> Else,
>
>> if the number of invalid blocks in the $3 \times 3$ neighborhood of $b \geq 5$,
>>
>>> mark $b$ as invalid.

4. *Level-3 detection*: For each valid block $b$ after Level-2 detection, we conduct the following:

> If $b$ is a block in the border of the image, then
>
>> if number of invalid blocks in the $5 \times 5$ neighborhood of $b \geq 8$,
>>
>>> mark $b$ as invalid.
>
> Else,
>
>> if number of invalid blocks in the $5 \times 5$ neighborhood of $b \geq 13$,
>>
>>> mark $b$ as invalid.

In general, if a block has been tampered with, then there are more chances that the neighboring blocks have also been tampered with. Thus, Level 2 and Level 3 reduce the false acceptance ratio of invalid blocks and helps to reduce the false alarms.

5. *Level-4 detection*: This level is to detect any changes in the five MSBs of each pixel of the valid blocks and to thwart VQ attacks. In this level we mark the block invalid only if there is a mismatch of the feature with both of the mapping blocks, so that false alarms can be reduced. For each valid block $b$ after Level-3 detection, we set the three LSBs of each pixel in block $b$ to zero. Compute the average intensity of block $b$, which is denoted as $avg\_b$. Determine the corresponding mapping partner block $x$-$y$ from the lookup table and perform the following task:

If $x$ is valid, then

    extract the 12-bit watermark from $x$.

    If $b \in \{$upper half of the image$\}$, then

        compare the five MSBs of $avg\_b$ against bits 1 to 5 of the extracted watermark.

    Else,

        compare the five MSBs of $avg\_b$ against bits 6 to 10 of the extracted watermark.

    If there is any mismatch in bit comparison, then mark $b$ as invalid.

    If $b$ is invalid, then

      if $y$ is valid, then

        extract the 12-bit watermark from $y$.

        If $b \in \{$upper half of the image$\}$, then

            compare the five MSBs of $avg\_b$ against bits 1 to 5 of the extracted watermark.

        Else,

            compare the five MSBs of $avg\_b$ against bits 6 to 10 of the extracted watermark.

        If there is any mismatch in bit comparison, then mark $b$ as invalid.

        Else, mark $b$ as valid.

Else, if $y$ is valid, then

    extract the 12-bit watermark from $y$.

    If $b \in \{$upper half of the image$\}$, then

        compare the five MSBs of $avg\_b$ against bits 1 to 5 of the extracted watermark.

    Else

        compare the five MSBs of $avg\_b$ against bits 6 to 10 of the extracted watermark.

    If there is any mismatch in bit comparison, then mark $b$ as invalid.

### 5.6.2.3 Recovery of Tampered Blocks

As in Lee and Lin [20], the recovery of tampered blocks is done in two stages. Stage 1 is responsible for recovery from the valid mapping blocks, which is the recovery from the embedded watermark. In Stage 2, we recover the image by averaging the $3 \times 3$ neighborhood.

### 5.6.2.3.1    Stage-1 Recovery

1. For each invalid block $b$, find the mapping block $x$ from the lookup table.
2. If $x$ is valid, then $x$ is the candidate block and go to step 5.
3. Determine the partner block of $x$ and denote it as $y$.
4. If $y$ is valid, then $y$ is the candidate block; otherwise, stop and leave $b$ as invalid.
5. Retrieve the 12-bit watermark mark from the candidate block.
6. If block $b$ is in the upper half of the image, then the 5-bit representative information of block $b$ starts from the first bit; otherwise, it starts from the sixth bit.
7. Pad three zeros to the end of 5-bit representative information.
8. Replace each pixel in block $b$ with this new 8-bit intensity and mark $b$ as valid.

### 5.6.2.3.2    Stage-2 Recovery

For each invalid block $b$, we recover the block by averaging the surrounding pixels of block $b$ in the $3 \times 3$ neighborhood, as shown in Figure 5.23.

### 5.6.2.4    Extension to Color Images

To extend the proposed scheme to color images, we consider the red, green, and blue components of the color image. The embedding process is done in each of the three components individually, and then each component is assembled to obtain a watermarked image, as shown in Figure 5.24a. Like the embedding process, the detection is also done on individual components. Then, each block in the three components is validated, such that if the block is invalid in any of the red, green, and blue components, then mark the block invalid in all three components. Finally, each component is reassembled to obtain the watermarked image, as shown in Figure 5.24b. The recovery process is conducted in each of the components individually, as described in Section 5.6.2.3, by considering the localization image in Figure 5.24b. Finally, combining the red, green, and blue components gives the recovered color image.

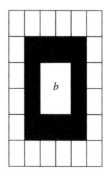

**FIGURE 5.23**    The $3 \times 3$ neighborhood of block $b$.

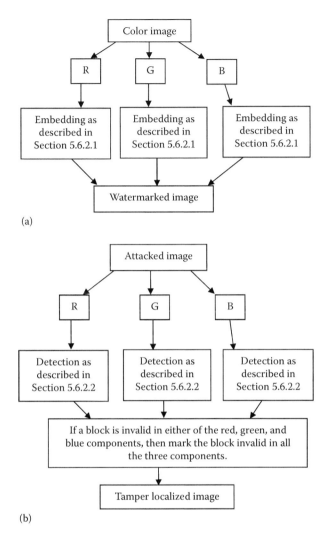

**FIGURE 5.24** (a) Embedding process for color images, (b) tamper detection process for color images.

### 5.6.3 Results and Discussions

In this section we use two different measures: one is the peak signal-to-noise ratio (PSNR), which measures the quality of the watermarked image, and the other is the percentage of recovery (PR), which compares the recovery rate of our technique with the existing techniques. The PSNR of an image $A$ with respect to an image $B$ is defined by

$$PSNR = 20 \times \log_{10} \left( \frac{MAX_I}{\sqrt{MSE}} \right), \tag{5.11}$$

where $MAX_I$ is the maximum gray value of the images $A$ and $B$.

$MSE$ is the mean square error defined by

$$MSE = \frac{1}{m \times n} \sum_{i=0}^{m-1} \sum_{j=0}^{n-1} \left\| A(i,j) - B(i,j) \right\|^2,$$ (5.12)

where $m \times n$ is the size of the images $A$ and $B$. The greater the PSNR, the lesser the distortion—that is, the greater the image quality. The $PR$ of an image $I$ is defined by

$$PR(I) = \left( \frac{SIZE_R(I)}{SIZE_M(I)} \right) \times 100,$$ (5.13)

where $SIZE_M(I)$ is the size of the modified portion in image $I$ and $SIZE_R(I)$ is the size of the recovered portion of the modified portion after Stage-1 recovery—that is, recovery from the features stored in the mapping block. The greater the $PR$, the more is recovered, and vice versa. Experiments are conducted on grayscale images of size $256 \times 256$ with $k = 13$ and 101 as the secret polynomial. Table 5.2 lists the PSNR values of different watermarked images by applying the proposed method.

We conduct various types of cropping attacks in the experiments. There are 12 distinct cropping patterns used, as shown in Figure 5.25. Table 5.3 lists the resulting $PR$ values from applying Lin et al. [17], Lin et al. [19], and the proposed schemes under these 12-pattern cropping attacks. We observe that the proposed algorithm outperforms the other two algorithms.

We also measure the performance of the proposed technique in terms of $PR$ by considering different percentages of cropping. The cropping is carried out from the top-left corner of the image. The graph in Figure 5.26 gives the performance of the proposed scheme against Lin et al. [17] and Lin et al. [19]. It is observed that the proposed scheme produces a much higher number of recovery pixels than the other two schemes. Figure 5.27 shows the application of our technique on a color image, where the license plate number of the car image was removed and can still be recovered. Figure 5.28 demonstrates that the use of random shape cropping cannot affect the accurate recovery of the Barbara image.

Lee and Lin [20] used the parity sum for authenticating individual blocks. This approach has the problem of detecting only even numbers of changes in the bit string. In their authentication method, the first authentication bit is the XOR of the first 10 bits of the feature, and the second authentication bit is the opposite of the first authentication bit. Therefore, if each pixel is modified with a gray value whose two LSBs are the opposites of each other, then it is less likely that their method

**TABLE 5.2**

**_PSNR_ Values of Different Watermarked Images by the Proposed Method**

|  | Lena | Baboon | Barbara | Cameraman | Peppers |
|---|---|---|---|---|---|
| _PSNR_ | 41.14 | 41.14 | 41.14 | 40.35 | 41.14 |

**FIGURE 5.25** Different cropping styles: (a) the original *Lena* $256 \times 256$ image, (b) the watermarked *Lena* image, (c) top 64: cropping $64 \times 256$ from the top, (d) bottom 64: cropping $64 \times 256$ from the bottom, (e) left 64: cropping $256 \times 64$ from the left, (f) right 64: cropping $256 \times 64$ from the right.

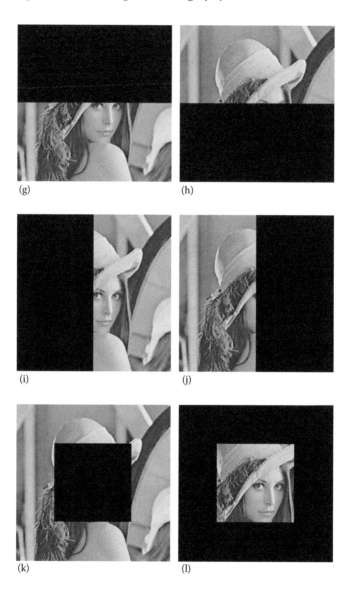

**FIGURE 5.25**    (Continued) Different cropping styles: (g) top 128: cropping $128 \times 256$ from the top, (h) bottom 128: cropping $128 \times 256$ from the bottom, (i) left 128: cropping $256 \times 128$ from the left, (j) right 128: cropping $256 \times 128$ from the right, (k) center: cropping $128 \times 128$ from the center, (l) outer: keeping the center $128 \times 128$.

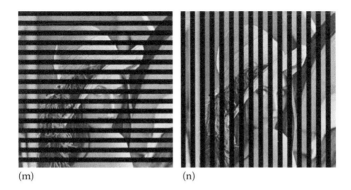

(m)         (n)

**FIGURE 5.25** (Continued) Different cropping styles: (m) horizontal: cropping the alternate $8 \times 256$ from the top, (n) vertical: cropping the alternate $256 \times 8$ from the left.

**TABLE 5.3**

**Comparisons of the *PR* Values under the 12-Pattern Cropping Attacks**

| Cropping style | Lin et al. [17] | Lin et al. [19] | Proposed |
|---|---|---|---|
| | PR | PR | PR |
| Top 64 | 75.00 | 69.34 | 100.00 |
| Bottom 64 | 75.00 | 69.34 | 100.00 |
| Left 64 | 75.00 | 62.50 | 100.00 |
| Right 64 | 75.00 | 62.50 | 100.00 |
| Top 128 | 50.00 | 46.24 | 100.00 |
| Bottom 128 | 50.00 | 46.24 | 100.00 |
| Left 128 | 50.00 | 40.63 | 100.00 |
| Right 128 | 50.00 | 40.63 | 100.00 |
| Outer 64 | 27.86 | 23.57 | 50.80 |
| Center 64 | 83.59 | 90.33 | 100.00 |
| Horizontal 8 | 50.00 | 46.09 | 45.90 |
| Vertical 8 | 50.00 | 50.00 | 48.44 |

will detect the modification. Figure 5.29 shows the results of the modification and its corresponding localization with a different gray value other than black. Their method detects modification in the case of black pixel replacement, but fails to detect the modification by replacing some other gray values. In Figure 5.29f, we show that the proposed method can detect both black pixels and other gray value modifications.

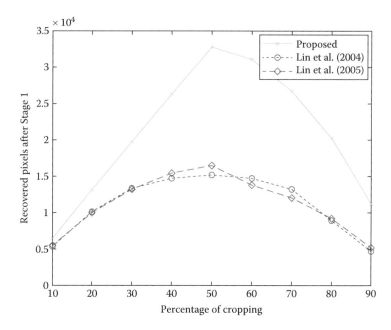

**FIGURE 5.26**    Comparisons of the proposed scheme with different percentages of cropping.

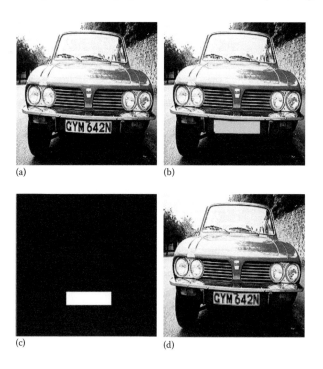

**FIGURE 5.27**    (a) A watermarked image containing a car, (b) a modified image with the license plate number removed, (c) the result of tamper detection, (d) the result of the recovered image.

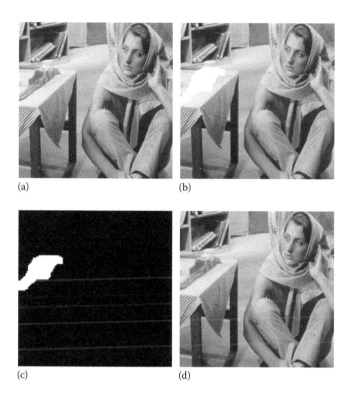

(a)                        (b)

(c)                        (d)

**FIGURE 5.28**   (a) A watermarked Barbara image, (b) a modified image with random shape cropping, (c) the result of the tamper localization image, (d) the result of the recovered image.

(a)                        (b)

**FIGURE 5.29**   (a) An original image, (b) the modification of the license plate by gray value 193.

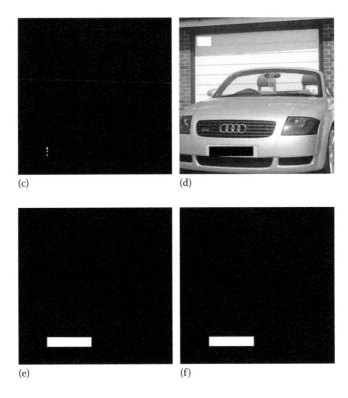

**FIGURE 5.29**    (Continued) (c) the detection of (b) by Lee and Lin, (d) the modification of the license plate by black value 0, (e) the detection of (d) by Lee and Lin, (f) the detection of (b) and (d) by our proposed scheme.

## REFERENCES

1. Cox, J. et al., A secure, robust watermark for multimedia, in *Proc. IEEE Int. Conf. Image Processing*, Lausanne, Switzerland, 1996, 185.
2. Kundur, D. and Hatzinakos, D., Diversity and attack characterization for improved robust watermarking, *IEEE Trans. Signal Processing*, 29, 2383, 2001.
3. Fei, C., Kundur, D., and Kwong, R. H., Analysis and design of secure watermark-based authentication systems, *IEEE Trans. Information Forensics and Security*, 1, 43, 2006.
4. Langelaar, G. C., Lagendijk, R. L., and Biemond, J., Removing spatial spread spectrum watermarks by non-linear filtering, in *Proc. Europ. Conf. Signal Processing*, Rhodes, Greece, 1998.
5. Pereira, S. et al., Second generation benchmarking and application oriented evaluation, in *Proc. Information Hiding Workshop III*, Pittsburgh, PA, 2001.
6. O'Ruanaidh, J. and Pun, T., Rotation, translation and scale invariant digital image Watermarking, in *Proc. Int. Conf. Image Processing*, 1997, 536.
7. Rosenfeld, A. and Kak, A. C., *Digital Picture Processing*, Academic Press, New York, 1982.
8. Petitcolas, F., Anderson, R., and Kuhn, M., Attacks on copyright marking systems, in *Proc. Int. Workshop on Information Hiding*, Portland, OR, 218, 1998.

9. Holliman, M. and Memon, N., Counterfeiting attacks for block-wise independent watermarking techniques, *IEEE Trans. Image Processing*, 9, 432, 2000.
10. Pereira, S. et al., Template based recovery of Fourier-based watermarks using logpolar and log-log maps, in *Proc. IEEE Int. Conf. Multimedia Computing and Systems*, Florence, Italy, 1999.
11. Linnartz, J. and Dijk, M., Analysis of the sensitivity attack against electronic watermarks in images, in *Proc. Int. Workshop on Information Hiding*, Portland, OR, 258, 1998.
12. Deguillaume, F. et al., Secure hybrid robust watermarking resistant against tampering and copy attack, *Signal Processing*, 83, 2133, 2003.
13. Kutter, M. and Petitcolas, F., A fair benchmark for image watermarking systems, in *Proc. SPIE Electronic Imaging 199: Security and Watermarking of Multimedia Content*, San Jose, CA, 1999, 25–27.
14. Petitcolas, F., Watermarking schemes evaluation, *IEEE Signal Processing*, 17, 58, 2000.
15. Voloshynovskiy, S. et al., Attack modelling: Towards a second generation benchmark, *Signal Processing*, 81, 1177, 2001.
16. Solachidis, V. et al., A benchmarking protocol for watermarking methods, in *Proc. IEEE Int. Conf. Image Processing*, Thessaloniki, Greece, 1023, 2001.
17. Lin, P. L. et al., A fragile watermarking scheme for image authentication with localization and recovery, in *Proc. IEEE Sixth Int. Symp. Multimedia Software Engineering*, Miami, FL, 146, 2004.
18. Tanebaum, A. S., *Computer Networks*, 4th edn., the Netherlands, Pearson Education International, 2003.
19. Lin, P. L. et al., A hierarchical digital watermarking method for image tamper detection and recovery, *Pattern Recognition*, 38, 2519, 2005.
20. Lee, T. Y. and Lin, S. D., Dual watermark for image tamper detection and recovery, *Pattern Recognition*, 41, 3497, 2008.

# 6 Combinational Domain Digital Watermarking

Digital watermarking plays a critical role in current state-of-the-art information hiding and security. It allows the embedding of an imperceptible watermark into multimedia data to identify ownership, trace authorized users, and detect malicious attacks. Research shows that the embedding of a watermark into the *least-significant bits* (LSBs) or the low-frequency components is reliable and robust. Therefore, we can categorize digital watermarking techniques into two domains: the spatial domain and the frequency domain.

In the spatial domain, one can simply insert a watermark into a host image by changing the gray levels of some pixels in the host image. The scheme is simple and easy to implement, but tends to be susceptible to attacks. In the frequency domain, we can insert a watermark into the coefficients of a transformed image—for example, using the *discrete Fourier transform* (DFT), *discrete cosine transform* (DCT), and *discrete wavelet transform* (DWT). This scheme is generally considered to be robust against attacks. However, one cannot embed a high-capacity watermark in the frequency domain since the image quality after watermarking would be degraded significantly.

To provide high-capacity watermarks and minimize image distortion, we present the technique of combining the spatial and frequency domains. The idea is to split the watermarked image into two parts that are respectively used for spatial and frequency insertions, depending on the user's preference and the importance of the data. The splitting strategy can be designed to be so complicated as to be insurmountable. Furthermore, to enhance robustness, a random permutation of the watermark is used to defeat image-processing attacks such as cropping.

This chapter is organized as follows. In Section 6.1 we present an overview of combinational watermarking. Sections 6.2 and 6.3 introduce the techniques of embedding watermarks in the spatial and frequency domains, respectively. We show some experimental results in Section 6.4. Section 6.5 discusses the further encryption of combinational watermarking.

## 6.1 OVERVIEW OF COMBINATIONAL WATERMARKING

In order to insert more data into a host image, the simplest way is to embed them in the spatial domain. However, the disadvantage is that the inserted data could be detected by some simple extraction skills. How can we insert more signals but keep the visual effects unperceivable? We present a new strategy of embedding a high-capacity watermark into a host image by splitting the watermark image into two parts: one is embedded in the spatial domain and the other in the frequency domain.

Let $H$ be the original gray-level host image of size $N \times N$ and $W$ be the binary watermark image of size $M \times M$. Let $W^1$ and $W^2$ denote the two separated watermarks from $W$, and let $H^S$ denote the combined image of $H$ and $W^1$ in the spatial domain. $H^{DCT}$ is the image once $H^S$ is transformed into the frequency domain by the DCT. $H^F$ is the image once $H^{DCT}$ and $W^2$ are combined in the frequency domain. Let $\oplus$ denote the operation that substitutes bits of the watermark for the LSBs of the host image.

The algorithm of the combinational image watermarking is presented in the following, and its flowchart is shown in Figure 6.1.

*Algorithm*:

1. Separate the watermark into two parts:

$$W = \{w(i, j),\ 0 \le i,\ j < M\},$$

where $w(i,j) \in \{0,1\}$.

$$W^1 = \{w^1(i, j),\ 0 \le i,\ j < M_1\},$$

where $w^1(i,j) \in \{0,1\}$.

$$W^2 = \{w^2(i, j),\ 0 \le i,\ j < M_2\},$$

where $w^2(i,j) \in \{0,1\}$.

$$M = M_1 + M_2.$$

2. Insert $W^1$ into the spatial domain of $H$ to obtain $H^S$ as

$$H^S = \{h^S(i, j) = h(i, j) \oplus w^1(i, j),\ 0 \le i,\ j < N\},$$

where $h(i,j)$ and $h^S(i,j) \in \{0, 1, 2, \ldots, 2^L-1\}$, and $L$ is the number of bits used as in the gray level of pixels.
3. Transform $H^S$ with the DCT to obtain $H^{DCT}$.

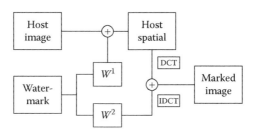

**FIGURE 6.1**     A flowchart in the combinational spatial and frequency domains.

4. Insert $W^2$ into the coefficients of $H^{DCT}$ to obtain $H^F$ as

$$H^F = \{h^F(i,j) = h^{DCT}(i,j) \oplus w^2(i,j), \ 0 \le i, j < N\},$$

where $h^F(i,j) \in \{0, 1, 2,\ldots, 2^L-1\}$ and $h^{DCT}(i,j) \in \{0, 1, 2,\ldots, 2^L-1\}$.

5. Transform the embedded host image with the inverse DCT.

The criteria of splitting the watermark image into two parts, which are individually inserted into the input image in the spatial and frequency domains, depend on user's requirements and applications. In principle, the most important information appears in the center of an image. Therefore, a simple way of splitting is to select the central window in the watermark image to be inserted into the frequency domain. With the user's preference, we can crop the most private data to be inserted into the frequency domain.

## 6.2 WATERMARKING IN THE SPATIAL DOMAIN

There are many ways of embedding a watermark into the spatial domain of a host image—for example, substituting the LSBs of some pixels [1], changing the paired pixels [2], and coding by textured blocks [3]. The most straightforward method is LSB substitution. Given a sufficiently high channel capacity in data transmission, a smaller object may be embedded multiple times. So, even if most of these watermarks are lost or damaged due to malicious attacks, a single surviving watermark would be considered a success.

Despite its simplicity, LSB substitution suffers a major drawback: any noise addition or lossy compression is likely to defeat the watermark. Even a simple attack is to simply set the LSB bits of each pixel to one. Moreover, the LSB insertion can be altered by an attacker without noticeable changes. An improved method would be to apply a pseudorandom number generator to determine the pixels to be used for embedding based on a designed key.

As shown in Figure 6.2, the watermarking can be implemented by modifying the bits of some pixels in the host image. Let $H^*$ be the watermarked image. The algorithm is presented as follows.

*Algorithm*:

1. Obtain the pixels from the host image.

$$H = \{h(i,j), 0 \le i, j < N\},$$

where $h(i,j) \in \{0, 1, 2,\ldots, 2^L-1\}$.

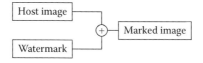

**FIGURE 6.2**  A flowchart in the spatial domain.

2. Obtain the pixels from the watermark.

$$W = \{w(i, j),\ 0 \le i,\ j < M\}.$$

3. Substitute the pixels of the watermark into the LSB pixels of the host image.

$$H^* = \left\{ h^*(i, j) = h(i, j) \oplus w(i, j), 0 <= i, j < N \right\},$$

where $h^*(i,j) \in \{0, 1, 2, \ldots, 2^L-1\}$.

## 6.3  WATERMARKING IN THE FREQUENCY DOMAIN

Watermarking can be applied in the frequency domain by first applying a transform. Similar to spatial domain watermarking, the values of specific frequencies can be modified from the original. Since high-frequency components are often lost by compression or scaling, the watermark is embedded into lower-frequency components, or alternatively, embedded adaptively into the frequency components that include critical information about the image. Since the watermarks applied to the frequency domain will be dispersed everywhere in the spatial image after the inverse transform, this method is not as susceptible to cropping attacks as the spatial domain approaches.

Several approaches can be taken to frequency domain watermarking—for example, the JPEG-based [4], spread spectrum [5,6], and content-based approaches [7]. Frequently used transformation functions are the DCT, DWT, and DFT. They allow an image to be separated into different frequency bands, making it much easier to embed a watermark by selecting the frequency bands of an image. One can avoid the most important visual information in the image (i.e., low frequencies) without making themselves noticeable for removal through compression and noise attacks (i.e., high frequencies).

Generally, we can insert the watermark into the coefficients of a transformed image as shown in Figures 6.3 and 6.4. The important consideration is what locations are best for embedding the watermark in the frequency domain to avoid distortion [8].

Let $H^m$ and $W^n$ be the subdivided images from $H$ and $W$, respectively, $H^{m\_DCT}$ be the image transformed from $H^m$ by the DCT, and $H^{m\_F}$ be the combined image of $H^{m\_DCT}$ and $W^n$ in the frequency domain. The algorithm is described as follows.

*Algorithm*:

1. Divide the host image into a set of $8 \times 8$ blocks.

$$H = \{h(i, j),\ 0 \le i,\ j < N\},$$

$$H^m = \{h^m(i, j),\ 0 \le i,\ j < 8\},$$

where $h^m(i,j) \in \{0, 1, 2, \ldots, 2^L-1\}$ and $m$ is the total number of the $8 \times 8$ blocks.

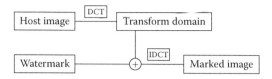

**FIGURE 6.3** A flowchart in the frequency domain.

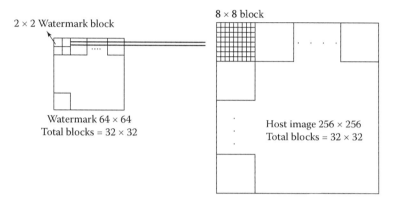

**FIGURE 6.4** The embedding skill in the frequency domain.

2. Divide the watermark image into a set of $2 \times 2$ blocks.

$$W = \{w(i, j),\ 0 \leq i,\ j < M\},$$

$$W^n = \{w^n(i, j),\ 0 \leq i,\ j < 2\},$$

where $w^n(i,j) \in \{0,1\}$ and $n$ is the total number of the $2 \times 2$ blocks.

3. Transform $H^m$ to $H^{m\_DCT}$ with the DCT.
4. Insert $W^m$ into the coefficients of $H^{m\_DCT}$.

$$H^{m-F} = \{h^{m-F}(i, j) = h^{m\_DCT}(i, j) \oplus w^m(i, j),\ 0 \leq i, j < 8\},$$

where $h^{m\_DCT}(i,j) \in \{0, 1, 2,\ldots, 2^L{-}1\}$.

5. Transform the embedded host image, $H^{m-F}$, with the inverse DCT.

The criterion for embedding the watermark image into the frequency domain of a host image is that the total number of $8 \times 8$ blocks in the host image must be larger than the total number of $2 \times 2$ blocks in the watermark image.

## 6.4   EXPERIMENTAL RESULTS

Figure 6.5 illustrates that it is important to split some parts of a watermark for security purposes. For example, people cannot view who the writer is in the images. Therefore, it is embedded into the frequency domain and the rest is embedded into the spatial domain. In this way, we not only enlarge the capacity but also secure the information concerned.

Figure 6.6 shows an original Lena image of size $256 \times 256$. Figure 6.7 is the traditional watermarking technique of embedding a $64 \times 64$ watermark into the frequency domain of the host image. In Figure 6.7, (a) is the original $64 \times 64$ watermark image and (b) is the watermarked Lena image given by embedding (a) into the frequency domain of Figure 6.6. Figure 6.7c is the extracted watermark image from Figure 6.7b.

Figure 6.8 demonstrates the embedding of a large watermark, a $128 \times 128$ image, into a host image. In Figure 6.8, (a) is the original $128 \times 128$ watermark image, and (b) and (c) are the two divided images from the original watermark, respectively. In Figure 6.8, we obtain (e) by embedding (b) into the spatial domain of the Lena image

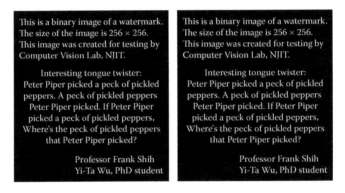

**FIGURE 6.5**    A $64 \times 64$ square area cut from a $256 \times 256$ image.

**FIGURE 6.6**    A Lena image.

**FIGURE 6.7** The traditional technique of embedding a $64 \times 64$ watermark into a Lena image.

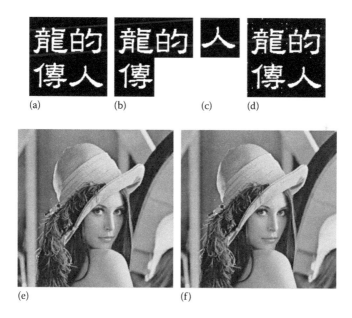

**FIGURE 6.8** The results of embedding a $128 \times 128$ watermark into a Lena image.

and obtain (f) by embedding (c) into the frequency domain of the watermarked Lena image in (e). Figure 6.8d is the extracted watermark from Figure 6.8f.

Figure 6.9a shows a larger watermark, a $256 \times 256$ image, split into two parts, as in Figure 6.5. Figures 6.9c,d are the watermarked images after embedding the watermark in the spatial and frequency domains of a host image, respectively. Figure 6.9b is the extracted watermark from Figure 6.9d.

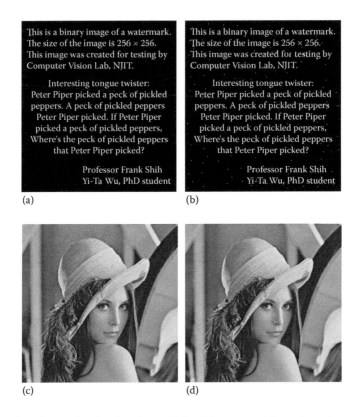

**FIGURE 6.9**   The results of embedding a 256 × 256 watermark into a Lena image.

We can apply error measures such as *normalized correlation* (NC) and *peak signal-to-noise ratio* (PSNR) to compute the image distortion after watermarking. The correlation between two images is often used in feature detection. Normalized correlation can be used to locate a pattern on a target image that best matches the specified reference pattern from the registered image base. Let $h(i,j)$ denote the original image and $h^*(i,j)$ denote the modified image. The normalized correlation is defined as

$$NC = \frac{\sum_{i=1}^{N}\sum_{j=1}^{N} h(i,j)h^*(i,j)}{\sum_{i=1}^{N}\sum_{j=1}^{N}\left[h(i,j)\right]^2}.$$

The PSNR is often used in engineering to measure the signal ratio between the maximum power and the power of the corrupting noise. Because signals possess a largely wide dynamic range, we apply the logarithmic decibel scale to limit its variation. It can measure the quality of reconstruction in image compression. However, it is a rough quality measure. In comparing two video files, we can calculate the mean PSNR. The peak signal-to-noise ratio is defined as

**TABLE 6.1**

**Comparison of Variously Sized Watermarks**
**Embedded into a Lena Image of Size 256 × 256**

|  | **64 × 64** | **128 × 128** | **256 × 256** |
|---|---|---|---|
| PSNR, first step | None | 56.58 | 51.14 |
| PSNR, second step | 64.57 | 55.93 | 50.98 |
| NC | 1 | 0.9813 | 0.9644 |

$$PSNR = 10\log_{10}\left(\frac{\displaystyle\sum_{i=1}^{N}\sum_{j=1}^{N}\left[h^{*}(i,j)\right]^{2}}{\displaystyle\sum_{i=1}^{N}\sum_{j=1}^{N}\left[h(i,j)-h^{*}(i,j)\right]^{2}}\right).$$

Table 6.1 presents the results of embedding different sizes of watermark into a host image. The PSNR in the first step compares both the original and the embedded Lena images in the spatial domain. The PSNR in the second step compares both images in the frequency domain. The NC compares both the original and the extracted watermarks.

## 6.5   FURTHER ENCRYPTION OF COMBINATIONAL WATERMARKING

For the purpose of enhancing robustness, a random permutation of the watermark is used to defeat image-processing attacks such as cropping. The procedure is illustrated in Figure 6.10.

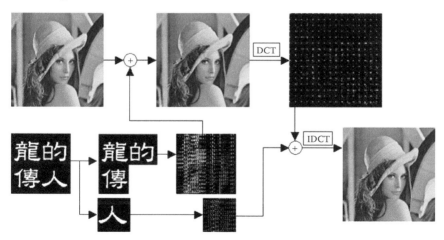

**FIGURE 6.10.**   The procedures of randomly permuting the watermark.

A random sequence generator is used to relocate the order of sequential numbers [9]. For example, the 12-bit random sequence generator is used to relocate the order of a watermark of size $64 \times 64$, as shown in Figure 6.11. We rearrange bits 9 and 6 to the rear of the whole sequence, and the result is shown in Figure 6.11b.

Figure 6.12 shows the result when half of the Lena image is cropped, where (a) is the original $128 \times 128$ watermark, (b) is the cropped Lena image, and (c) is the extracted watermark.

Table 6.2 shows the results when parts of an embedded host image with a watermark of $128 \times 128$ are cropped. For example, if a $64 \times 64$ section is cropped from a $256 \times 256$ embedded host image, the NC is 0.92. If half of the embedded host image is cropped (i.e., eight sections of $64 \times 64$), as shown in Figure 6.12b, the NC is 0.52.

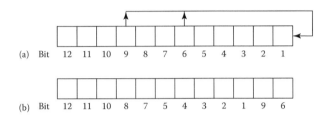

**FIGURE 6.11**    A 12-bit random sequence generator.

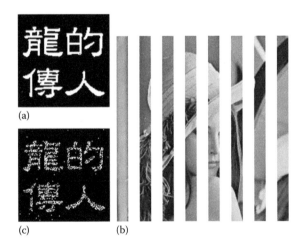

**FIGURE 6.12**    The results when cropping half of a Lena image.

**TABLE 6.2**

**Comparison of a Lena Image Cropped by Multiples of 16 × 256 Stripes**

| Multiples of 16 × 256 | 1 | 2 | 3 | 4 | 5 | 6 | 7 | 8 |
|---|---|---|---|---|---|---|---|---|
| NC | 0.92 | 0.88 | 0.81 | 0.73 | 0.68 | 0.61 | 0.58 | 0.52 |

Generally speaking, the important part of an image is not enormous. We can cut out the important part and embed it into the frequency domain and the rest into the spatial domain of a host image. Therefore, we can not only enlarge the size of the watermark, but also maintain the properties of security and imperceptibility. Combinational image watermarking possesses the following advantages [10]: more watermark data can be inserted into the host image, so the capacity is increased; the splitting of the watermark into two parts doubles the degree of protection; and the splitting strategy can be designed to be so complicated as to be insurmountable.

Another scheme of combining spatial domain watermarking with wavelet domain watermarking can be referred to [11]. Employing error control codes could increase the robustness of spatial domain watermarking [12], but the watermark capacity is reduced since the codes add some redundancy to the watermark.

## REFERENCES

1. Wolfgang, R. and Delp, E., A watermarking technique for digital imagery: Further studies, in *Proc. Int. Conf. Imaging Science, Systems and Technology*, Las Vegas, NV, pp. 279–287, 1997.
2. Pitas, I. and Kaskalis, T., Applying signatures to digital images, in *Proc. IEEE Workshop Nonlinear Signal and Image Processing*, Halkidiki, Greece, June 1995, 460.
3. Caronni, G., Assuring ownership rights for digital images, in *Proc. Reliable IT Systems*, Vieweg, Germany, pp. 251–263, 1995.
4. Zhao, K. E., Embedding robust labels into images for copyright protection, Technical Report, Fraunhofer Institute for Computer Graphics, Darmstadt, Germany, 1994.
5. Cox, I. et al., Secure spread spectrum watermarking for images audio and video, in *Proc. IEEE Int. Conf. Image Processing*, vol. 3, Lausanne, Switzerland, 1996, 243.
6. Cox, I. et al., Secure spread spectrum watermarking for multimedia, *IEEE Trans. Image Processing*, 6, 1673, 1997.
7. Bas, P., Chassery, J.-M., and Macq, B., Image watermarking: An evolution to content based approaches, *Pattern Recognition*, 35, 545, 2002.
8. Hsu, C.-T. and Wu, J.-L., Hidden signature in images, *IEEE Trans. Image Processing*, 8, 58, 1999.
9. Lin, D. and Chen, C.-F., A robust DCT-based watermarking for copyright protection, *IEEE Trans. Consumer Electronics*, 46, 415, 2000.
10. Shih, F. Y. and Wu, S. Y. T., Combinational image watermarking in the spatial and frequency domains, *Pattern Recognition*, 36, 969, 2003.
11. Tsai, M. J., Yu, K. Y., and Chen, Y. Z., Joint wavelet and spatial transformation for digital watermarking, *IEEE Trans. Consumer Electronics*, 46, 241, 2000.
12. Lancini, R., Mapelli, F., and Tubaro, S., A robust video watermarking technique for compression and transcoding processing, in *Proc. IEEE Int. Conf. Multimedia and Expo*, Zadar, Croatia, 2002, 549.

# 7 Watermarking Based on Genetic Algorithms

In digital watermarking, the substitution approach is a popular method for embedding watermarks into an image. In order to minimize image quality degradation, *least-significant-bit* (LSB) substitution is commonly used. In general, LSB substitution works well with cryptographic technology such as message digest algorithm 5 (MD5) and the Rivest–Shamir–Adleman (RSA) algorithm in the spatial domain [1,2]. However, LSB substitution may fail in the frequency domain due to rounding errors.

As illustrated in Figure 7.1, the watermark is embedded into the *discrete cosine transform* (DCT) coefficients of a cover image to generate a frequency domain watermarked image by using the substitution embedding approach. After performing an *inverse DCT* (IDCT) on the frequency domain watermarked image, a spatial domain watermarked image, called the *stego-image*, can be obtained in which most values are real numbers. Since the pixel intensity is stored in the integer format, the real numbers have to be converted into integers. In the literature, a rounding operation is usually recommended to convert the real numbers to integers [3]. However, a question is raised as to whether the watermark can be correctly extracted from the stego-image. Unfortunately, the extracted watermark appears as a noisy image. Therefore, it is difficult to correctly retrieve the embedded data if the rounding technique is adopted as the substitution approach, especially LSB substitution, in the frequency domain.

In this chapter, we first briefly introduce *genetic algorithms* (GAs) in Section 7.1. We present the concept of GA-based watermarking in Section 7.2. The GA-based rounding error watermarking scheme is described in Section 7.3. We then explore the application of GA methodology to medical image watermarking in Section 7.4. The authentication of JPEG images based on GAs is presented in Section 7.5.

## 7.1 INTRODUCTION TO GENETIC ALGORITHMS

GAs, introduced by Hollard [4] in his seminal work, are commonly used as adaptive approaches to provide randomized, parallel, and global searching based on the mechanics of natural selection and natural genetics in order to find solutions to a problem. There are four distinctions from normal optimization and search procedures [5]: (1) GAs work with a coded parameter set, not the parameters themselves; (2) GAs search from randomly selected points, not from a single point; (3) GAs use objective function information; (4) GAs use probabilistic transition rules, not deterministic ones.

Although GAs have been developed in many ways [6,7], the fundamentals of GAs are based on the *simple genetic algorithm* (SGA) [8]. In general, GAs start with a

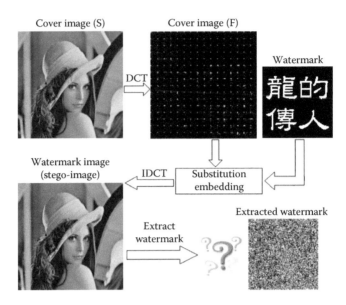

**FIGURE 7.1** An example of the rounding error problem caused by the substitution embedding approach in the frequency domain.

randomly selected population, called the *first generation*. Each individual, called a *chromosome*, in the population corresponds to a solution in the problem domain. An objective, called the *fitness function*, is used to evaluate the quality of each chromosome. The next generation will be generated from chromosomes whose fitness values are high. Reproduction, crossover, and mutation are the three basic operators used to reproduce chromosomes in order to obtain the best solution. The process will repeat many times until a predefined condition is satisfied or the desired number of iterations is reached. A detailed description of chromosomes, operations, and fitness functions will be presented in the following sections.

### 7.1.1 CHROMOSOMES

If a problem is considered a lock then a chromosome can be considered a key. That is, each chromosome can be treated as a possible solution to a problem. Therefore, GAs can be thought of as the procedure of selecting the correct key to open a lock, as shown in Figure 7.2.

Usually, there are no matches among the existing keys; that is, no key can be used to open the lock. In this case, the locksmith, using his experience, will select a candidate key and adjust its shape to generate the correct one to open the lock. Note that the criterion of selecting the candidate key is based on the adjustment complexity; that is, the smaller the adjustment the candidate key requires, the higher the suitability of the candidate key. The same logic can be applied to GAs. The chromosomes of the new generation are obtained based on the high-quality chromosomes of the previous generation. The procedure of evaluating a chromosome will be presented in Section 7.1.3. An example of generating the desired chromosome using GA operations is shown in

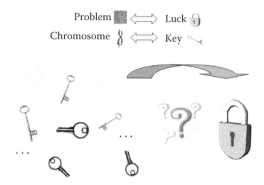

**FIGURE 7.2**   The procedure of selecting the correct key to open a lock.

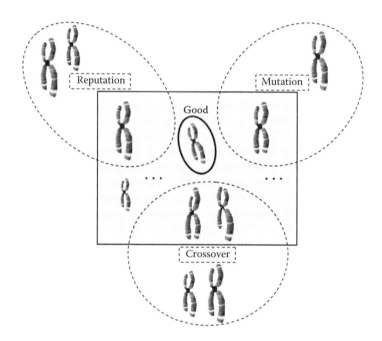

**FIGURE 7.3**   An example of generating the desired chromosome with GA operations.

Figure 7.3. The three GA operations—reproduction, crossover, and mutation—are used to adjust candidate chromosomes and generate a new chromosome. Note that the procedure of creating the new generation will not be terminated until the desired chromosome is obtained.

## 7.1.2   Basic Operations of Genetic Algorithms

The basic operations of GAs—reproduction, crossover, and mutation—are described in the following sections.

### 7.1.2.1   Reproduction

The purpose of reproduction is to avoid blindly searching for the desired solution by ensuring the high quality of chromosomes; that is, they are much closer to the ideal chromosomes in the current generation to be maintained in the next generation. Figure 7.4 shows an example of the reproduction operation. Let the number of chromosomes in the current generation be eight and the maintaining ratio 0.25 (two chromosomes will be maintained among eight chromosomes). Suppose the requirement of the ideal chromosome is that the genes of a chromosome should belong to the same gene (same size and same shape). It is obvious that the two chromosomes are obtained after reproduction since each of them contains different genes.

### 7.1.2.2   Crossover

The purpose of crossover is to avoid a situation where the searching procedure is blocked on local minimum/maximum subsets by significantly modifying the genes of a chromosome. Figure 7.5 shows an example of the crossover operation. The procedure of crossover starts by randomly selecting two chromosomes as parents from the current generation, followed by the exchange of genes of the two parent chromosomes. Let the requirement of the ideal chromosome be the same as that shown in the reproduction operation. It is obvious that the two new chromosomes obtained by the crossover operation are close to the requirement of the ideal chromosome.

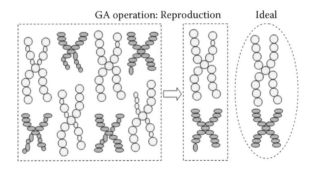

**FIGURE 7.4**   An example of reproduction.

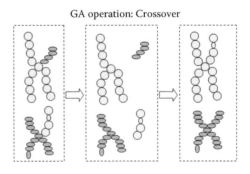

**FIGURE 7.5**   An example of crossover.

### 7.1.2.3 Mutation

The purpose of mutation is to provide a local minimum/maximum searching approach by slightly modifying the gene of a chromosome. The mutation operation works well especially in cases where the difference between a chromosome and the desired one is very small. For example, in Figure 7.6, the difference between the two chromosomes and the desired ones is 2. After the mutation operation, the four chromosomes obtained as the new generation are close to the desired ones, since the number of differences is decreased to 1.

### 7.1.3 FITNESS FUNCTION

The fitness function is used to evaluate the quality of the chromosomes in the current generation. Usually, the fitness function utilizes the threshold value to distinguish the good chromosomes from the bad ones and is designed based on the problem. For example, based on the requirement as shown in the reproduction operation, the fitness function will be defined as follows: the number of genes that do not belong to the dominant gene. The procedure to determine the dominant gene is as follows. First, the same kinds of genes among a chromosome are collected together. Secondly, the numbers of genes of each category are obtained. Thirdly, the gene with largest number will be determined as the dominant gene. Figure 7.7 shows an example

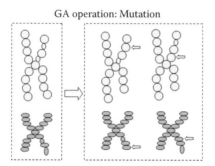

**FIGURE 7.6**   An example of mutation.

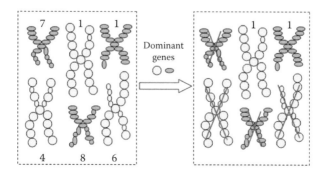

**FIGURE 7.7**   An example of fitness function.

of the fitness function. There are six chromosomes in the original generation. It is obvious there are two kinds of dominant gene among them. The value along each chromosome is the total number of nondominant genes of a chromosome. Now, let the threshold value be 2. Therefore, only two chromosomes can pass through the evaluation of the fitness function.

## 7.2    CONCEPT OF GA-BASED WATERMARKING

As previously mentioned, it is possible to retrieve a noisy watermark from a water-marked image if the LSB substitution approach is utilized as the embedding strategy and the watermark is embedded into the frequency domain. Therefore, is it possible to correct the error by changing the value of certain locations in the spatial domain? The answer is yes, but it is not easy, since it is difficult to predict what the impacts will be in the frequency domain of a cover image if changes are made to the values of pixels in the spatial domain.

In order to solve the problem, evolutionary algorithms such as simulated anneal-ing, hill climbing, and GAs are utilized. In simulated annealing, the choice of initial temperature and the corresponding temperature decrement strategy will affect the performance dramatically. The purpose of randomly selecting the neighbors of a candidate in simulated annealing is to find the optimum solution. However, in the rounding error problem, it is not easy to decide all of them since there is no relation between a pixel and its neighbors. In the hill-climbing method, two main procedures are used to determine the initial states and the next states. Since the relation between an image and its frequency domain is complicated, it is not easy to find the next state. Therefore, GAs are considered to be the best method for solving the rounding error problem.

Before presenting the basic concept of reducing errors with GAs, we consider the following two primary issues:

1. The embedded data should be accurate.
2. The changes made in order to modulate the errors should be minimal.

Note that the coefficients in the frequency domain will be changed dramatically even though only one pixel in the spatial domain is changed. Moreover, changes to pixels in the frequency domain are difficult to derive intuitively. Therefore, it is dif-ficult to determine whether the embedded data are variant or not in the process of translating real numbers into integers. On the other hand, it is also difficult predict and change the proper pixels in order to correct the errors and then obtain the exact embedded data. Therefore, it is not only difficult to modulate intuitively, but also likely to cause the worst result by random modulation.

In order to solve the preceding problems, a GA is utilized to find a suitable solu-tion for translating the real numbers into integers instead of the traditional rounding approach, as shown in Figure 7.8. The GA-based novel method not only restores the embedded data exactly, but also changes the least amount of pixels in the cover image.

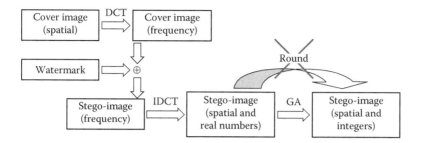

**FIGURE 7.8** The purpose of a GA.

## 7.3 GA-BASED ROUNDING ERROR CORRECTION WATERMARKING

In the literature, a simple rounding strategy is usually recommended for converting real numbers into integers. However, rounding errors, as mentioned earlier, will cause problems such that the embedded watermark is significantly destroyed. An illustration of the rounding error problem is shown in Figure 7.9.

In Figure 7.9, (a) is the original host image, an $8 \times 8$ gray-level image, in the spatial domain and (b) is the image transformed by the DCT; (c) is a binary watermark, in which 0 and 1 denote the embedded value in its location and a dash (–) indicates no change in its position; (d) is obtained by embedding (c) into (b) based on LSB modification. Note that the watermark is embedded into the integer part of an absolute real number. After transforming (d) into its spatial domain by the IDCT, the watermarked (stego-) image is generated, as shown in (e), where all pixels are real numbers. After rounding the real numbers into integers in (e), the watermarked image is obtained in (f).

The method of extracting the watermark is as follows. The watermark can be extracted from the coefficients of the frequency domain of the watermarked image (Figure 7.9f) by performing the embedding procedure again. Therefore, (f) is transformed by the DCT to obtain (g). Finally, the embedded watermark is obtained, as shown in (h), by extracting data from specific positions in (g). For example, the integer part of an absolute value of −47.94 is 47. The binary string 00101110 is obtained by translating the decimal value 46 into its binary format. Then, the value 0 is obtained from the LSB 00101110. Similarly, a 4-bit watermark (0,0,1,0) can be determined from (−47.94,30.87,−5.16,94.24). Comparing the embedded (1101) and extracted (0010) watermark, the unfortunate result is that the watermark is totally different. In other words, the original watermark is lost by the rounding approach.

Table 7.1 shows the results when embedding different kinds of watermarks into the same transformed image in Figure 7.9d. The embedded rule is shown in Figure 7.10. That is, for a 4-bit watermark 1234, we insert 1, the *most significant bit* (MSB), into position A, 2 into B, 3 into C, and 4 into D. There are $2^4$ possible embedded watermarks. Only two cases can be extracted correctly.

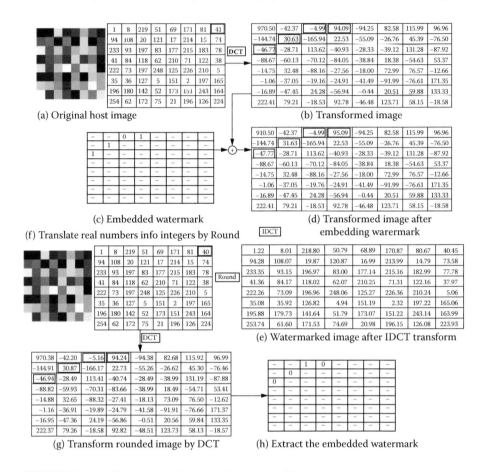

(a) Original host image

(b) Transformed image

(c) Embedded watermark

(d) Transformed image after embedding warermark

(f) Translate real numbers info integers by Round

(e) Watermarked image after IDCT transform

(g) Transform rounded image by DCT

(h) Extract the embedded watermark

**FIGURE 7.9** An illustration of the rounding error problem.

## TABLE 7.1
## Total Possible Embedded Watermarks

| Embedded | Extracted | Error Bits | Embedded | Extracted | Error Bits |
|----------|-----------|------------|----------|-----------|------------|
| 0000 | 0000 | 0 | 1000 | 0000 | 1 |
| 0001 | 0000 | 1 | 1001 | 0000 | 2 |
| 0010 | 0000 | 1 | 1010 | 0000 | 2 |
| 0011 | 0000 | 2 | 1011 | 0000 | 3 |
| 0100 | 0000 | 1 | 1100 | 0000 | 2 |
| 0101 | 0000 | 2 | 1101 | 0010 | 4 |
| 0110 | 0000 | 2 | 1110 | 1110 | 0 |
| 0111 | 1110 | 2 | 1111 | 1110 | 1 |

| | | C | D | | | | |
|---|---|---|---|---|---|---|---|
| – | – | C | D | – | – | – | – |
| – | B | – | – | – | – | – | – |
| A | – | – | – | – | – | – | – |
| – | – | – | – | – | – | – | – |
| – | – | – | – | – | – | – | – |
| – | – | – | – | – | – | – | – |
| – | – | – | – | – | – | – | – |

**FIGURE 7.10** The positions where the watermark is embedded.

### 7.3.1 DEFINITIONS OF CHROMOSOME, FITNESS FUNCTION, AND GA OPERATIONS

As mentioned earlier, a GA is utilized to determine the rules of translating real numbers into integers instead of the traditional rounding approach. The chromosomes, fitness functions, and GA operations used in the *GA-based rounding error correction* (GA-REC) algorithm are defined in the following subsections.

#### 7.3.1.1 Chromosomes

In order to apply GAs to our problem, a chromosome $G$ consisting of 64 genes is represented as $G = g_0, g_1, g_2,\ldots, g_{63}$, where $g_i (0 \leq i \leq 63)$ corresponds to the pixels shown in Figure 7.11. Figure 7.11a denotes the positions where $g_i (0 \leq i \leq 63)$ are located. For example, the chromosome $G_1$ shown in Figure 7.11b is represented as

$$G_1 = 1101010100010001001110010110111010101010100101011100101101111000.$$

The use of chromosomes in GA-based watermarking is different from traditional methods. That is, it is not necessary to decode each chromosome, the binary string, into a number for evaluating the fitness value. The chromosomes are used to represent the policy of translating real numbers into integers. Let $r$ be a real number. An integer $r^*$ can be obtained by the following rules:

1. If the signal is 1,

$$r^* = Trunc(r) + 1.$$

| 0 | 1 | 2 | 3 | 4 | 5 | 6 | 7 |
|---|---|---|---|---|---|---|---|
| 8 | 9 | 10 | 11 | 12 | 13 | 14 | 15 |
| 16 | 17 | 18 | 19 | 20 | 21 | 22 | 23 |
| 24 | 25 | 26 | 27 | 28 | 29 | 30 | 31 |
| 32 | 33 | 34 | 35 | 36 | 37 | 38 | 39 |
| 40 | 41 | 42 | 43 | 44 | 45 | 46 | 47 |
| 48 | 49 | 50 | 51 | 52 | 53 | 53 | 55 |
| 56 | 57 | 58 | 59 | 60 | 61 | 62 | 63 |

(a)

| 1 | 1 | 0 | 1 | 0 | 1 | 0 | 1 |
|---|---|---|---|---|---|---|---|
| 0 | 0 | 0 | 1 | 0 | 0 | 0 | 1 |
| 0 | 0 | 1 | 1 | 1 | 0 | 0 | 1 |
| 0 | 1 | 1 | 0 | 1 | 1 | 1 | 0 |
| 1 | 0 | 1 | 0 | 1 | 0 | 1 | 0 |
| 1 | 0 | 0 | 1 | 0 | 1 | 0 | 1 |
| 1 | 1 | 0 | 0 | 1 | 0 | 1 | 1 |
| 0 | 1 | 1 | 1 | 1 | 0 | 0 | 0 |

(b)

**FIGURE 7.11** The positions corresponding to the genes.

2. If the signal is 0,

$$r^* = Trunc(r),$$

where $Trunc(r)$ denotes the integer part of $r$.

An example is illustrated in Figure 7.12; (a) shows an image whose pixel values are real numbers; (b) is obtained by translating the real numbers into integers using the chromosomes in Figure 7.11b.

### 7.3.1.2   Fitness Function

It is obvious that certain numbers of chromosomes are created in each generation to provide possible solutions. In order to evaluate the quality of each chromosome (i.e. the correctness of extracting the desired watermarks from the corresponding solution), an evaluation function is designed to calculate the differences between the two watermarks, as shown in the following. Let $\xi$ be a chromosome, and $Watermark^\alpha$ and $Watermark^\beta$ be the embedded and extracted watermarks, respectively.

$$Evaluation\_1(\xi) = \sum_{i=0}^{\text{all pixels}} \left| Watermark^\alpha(i) - Watermark^\beta(i) \right|, \qquad (7.1)$$

where $Watermark^\alpha(i)$ and $Watermark^\beta(i)$ are the $i$th bit of the embedded and the extracted watermarks, respectively. Note that, if they are the same, $Evaluation\_1(\xi) = 0$.

### 7.3.1.3   Reproduction

A reproduction operation is defined as

$$Reproduction(\xi) = \{\xi_i, \xi_i \in \Psi, Evaluation(\xi_i) \leq \Omega\}, \qquad (7.2)$$

where $\Omega$, a critical value, is used to sieve chromosomes and $\Psi$ is the population. That is, we reproduce excellent chromosomes from the population because of their high quality. In our experience, we set $\Omega$ to be 10% of the total number of pixels.

| 1.22 | 8.01 | 218.80 | 50.79 | 68.89 | 170.87 | 80.67 | 40.45 |
|------|------|--------|-------|-------|--------|-------|-------|
| 94.28 | 108.07 | 19.87 | 120.87 | 16.99 | 213.99 | 14.79 | 73.58 |
| 233.35 | 93.15 | 196.97 | 83.00 | 177.14 | 215.16 | 182.99 | 77.78 |
| 41.36 | 84.17 | 118.02 | 62.07 | 210.25 | 71.31 | 122.16 | 37.97 |
| 222.26 | 73.09 | 196.96 | 248.06 | 125.27 | 226.36 | 210.24 | 5.06 |
| 35.08 | 35.92 | 126.82 | 4.94 | 151.19 | 2.32 | 197.22 | 165.06 |
| 195.88 | 179.73 | 141.64 | 51.79 | 173.07 | 151.22 | 243.14 | 163.99 |
| 253.74 | 61.60 | 171.53 | 74.69 | 20.98 | 196.15 | 126.08 | 223.93 |

(a)

| 2 | 9 | 218 | 51 | 68 | 171 | 80 | 41 |
|---|---|-----|----|----|-----|----|----|
| 94 | 108 | 19 | 121 | 16 | 213 | 14 | 74 |
| 233 | 93 | 197 | 84 | 178 | 215 | 182 | 78 |
| 41 | 85 | 119 | 62 | 211 | 72 | 123 | 37 |
| 223 | 73 | 197 | 248 | 126 | 226 | 211 | 5 |
| 36 | 35 | 126 | 5 | 151 | 3 | 197 | 166 |
| 196 | 180 | 141 | 51 | 174 | 151 | 244 | 164 |
| 253 | 62 | 172 | 75 | 21 | 196 | 126 | 223 |

(b)

**FIGURE 7.12**    The usage of the genes.

### 7.3.1.4 Crossover

A crossover operation is defined as

$$\text{Crossover}(\xi) = \{\xi_i \Theta \xi_j, \ \xi_i, \xi_j \in \Psi\}, \tag{7.3}$$

where $\Theta$ denotes the operation that recombines $\xi$ by exchanging genes from their parents, $\xi_i$ and $\xi_j$. This operation gestates a better offspring as they will inherit the best genes from their parents. The most often used crossovers are one-point, two-point, and multipoint crossovers. The best time to select a suitable crossover depends on the length and structure of the chromosomes. This can be illustrated clearly by an example, as shown in Figure 7.13.

### 7.3.1.5 Mutation

A mutation operation is defined as

$$\text{Mutation}(\xi) = \{\xi_i \circ j, \quad \text{where} \quad 0 \le j \le \text{Max\_Length}(\xi_i), \xi_i \in \Psi\}, \tag{7.4}$$

where $\circ$ is the operation of randomly selecting chromosome $\xi_i$ from $\Psi$ and randomly changing bits from $\xi_i$. Figure 7.14 shows the one-point and two-point mutations.

### 7.3.2 GA-BASED ROUNDING ERROR CORRECTION ALGORITHM

The GA-REC algorithm is presented as follows. Its flowchart is shown in Figure 7.15.

*GA-REC algorithm*:

1. Define the fitness function, number of genes, size of population, crossover rate, critical value, and mutation rate.
2. Create the first generation by random selection.

| | | One point | Point | | Two point | Point 1 | Point 2 |
|---|---|---|---|---|---|---|---|
| Before crossover | Chromosome A | 1011 | 010 | Chromosome A | 101111 | 010 | 0011 |
| | Chromosome B | 0010 | 111 | Chromosome B | 001011 | 111 | 1010 |
| After crossover | Chromosome A | 1011 | 011 | Chromosome A | 101111 | 111 | 0011 |
| | Chromosome B | 0010 | 010 | Chromosome B | 001011 | 010 | 1010 |

**FIGURE 7.13** One-point and two-point crossovers.

| | | One point | | Two point |
|---|---|---|---|---|
| Before mutation | Chromosome A | 1011010 ▲ | Chromosome A | 1011110100011 ▲  ▲ |
| After mutation | Chromosome A | 1010010 ▲ | Chromosome A | 1011010110011 ▲  ▲ |

**FIGURE 7.14** One-point and two-point mutations.

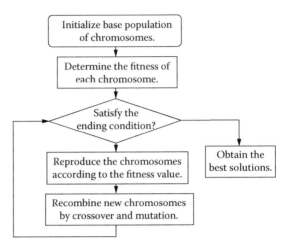

**FIGURE 7.15**    Basic steps to finding the best solutions by GAs.

3. Translate real numbers into integers based on each corresponding chromosome.
4. Evaluate the fitness value for each corresponding chromosome.
5. Obtain the best chromosomes based on their fitness values.
6. Recombine new chromosomes by using crossover.
7. Recombine new chromosomes by using mutation.
8. Repeat Steps 3 to 8 until a predefined condition is satisfied or the desired number of iterations is reached.

Figure 7.16 shows the result of error correction using a GA; (a) is the original cover image, an $8 \times 8$ gray-level image, in the spatial domain, and (b) is the original image transformed by the DCT; (c) is a binary watermark, in which 0 and 1 denote the embedded data in their location and a dash (–) indicates no change in their position; (d) is obtained by embedding (c) into (b) based on LSB modification. It is obvious that there are three differences when comparing (b) and (d)—for instance, (–47.77 and –47.77), (30.63 and 31.63), and (94.09 and 95.09). After transforming (d) into its spatial domain with the IDCT, we obtain (e), where all pixels are real numbers. After translating the real numbers into integers with a GA, (f) is generated; (g) is the transformed image from (f) by the DCT. Finally, the exact embedded watermark is obtained by extracting the bits from the same position of embedding the watermark, as shown in (h).

Figure 7.17 shows the other suitable solutions obtained by GAs. We transform (a) into (b) and (c) into (d), respectively. Although there are differences between (b) and (d), the watermark can be exactly extracted from the same position, as shown in (c).

### 7.3.3  ADVANCED STRATEGY FOR INITIALIZING THE FIRST POPULATION

In Figure 7.17, although numerous suitable solutions can be obtained by GA-REC to correct the errors, many of them are useless considering that the changes in the

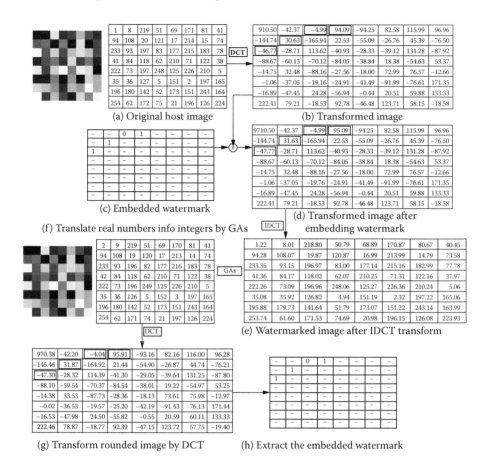

**FIGURE 7.16**  An example GA-REC.

original image should be as small as possible. That is, the *peak signal-to-noise ration* (PSNR) should be as high as possible. The definition of PSNR is

$$PSNR = 10 \times \log_{10} \left( \frac{255^2}{\displaystyle\sum_{i=1}^{N}\sum_{j=1}^{N}\left[h(i,j) - h^{GA}(i,j)\right]^2} \right). \quad (7.5)$$

Therefore, an improved method for minimizing the changes is developed. In order to limit changes, the first population is generated based on an initial chromosome $G_{Ini}$, which is generated by comparing the difference between the original and rounded images. For example, there is only one difference between Figures 7.18a and 7.18c, so $G_{Ini}$ is generated as follows:

$G_{Ini} = 00000001000000000000000000000000000000000000000000000000000000000.$

| 2 | 8 | 219 | 50 | 69 | 170 | 81 | 41 |
|---|---|---|---|---|---|---|---|
| 94 | 108 | 20 | 120 | 17 | 214 | 14 | 74 |
| 234 | 93 | 196 | 82 | 177 | 216 | 182 | 78 |
| 41 | 85 | 118 | 62 | 211 | 71 | 123 | 37 |
| 222 | 74 | 196 | 248 | 126 | 226 | 210 | 5 |
| 35 | 36 | 127 | 5 | 152 | 2 | 197 | 165 |
| 196 | 180 | 142 | 51 | 174 | 151 | 243 | 163 |
| 254 | 61 | 172 | 74 | 21 | 196 | 127 | 223 |

(a)

| 970.25 | -41.72 | -4.72 | 95.99 | -94.50 | 81.33 | 115.34 | 98.20 |
|---|---|---|---|---|---|---|---|
| -144.70 | 31.52 | -164.71 | 22.18 | -54.70 | -27.61 | 45.75 | -77.27 |
| -47.76 | -28.53 | 114.42 | -41.24 | -28.69 | -38.49 | 132.11 | -86.98 |
| -88.10 | -59.77 | -70.10 | -84.95 | -38.29 | 17.80 | -55.40 | 53.56 |
| -14.25 | 32.59 | -87.95 | -27.79 | -18.50 | 73.65 | 75.12 | -12.31 |
| -0.61 | -36.46 | -19.83 | -24.60 | -41.90 | -91.88 | -77.15 | 171.89 |
| -16.73 | -47.62 | 24.86 | -56.69 | -0.02 | 21.89 | 60.33 | 133.01 |
| 222.11 | 79.33 | -17.89 | 93.45 | -45.98 | 123.08 | 58.38 | -19.19 |

(b)

| 2 | 8 | 219 | 50 | 69 | 170 | 81 | 41 |
|---|---|---|---|---|---|---|---|
| 94 | 108 | 20 | 120 | 16 | 214 | 14 | 74 |
| 233 | 93 | 196 | 82 | 177 | 216 | 182 | 78 |
| 41 | 85 | 118 | 62 | 211 | 71 | 123 | 37 |
| 222 | 74 | 196 | 248 | 126 | 226 | 210 | 5 |
| 35 | 36 | 127 | 5 | 152 | 2 | 197 | 165 |
| 196 | 180 | 142 | 51 | 174 | 151 | 243 | 163 |
| 254 | 61 | 172 | 74 | 21 | 196 | 127 | 223 |

(c)

| 969.88 | -41.98 | -4.69 | 95.96 | -94.38 | 81.82 | 115.73 | 98.22 |
|---|---|---|---|---|---|---|---|
| -144.92 | 31.47 | -164.56 | 22.08 | -54.77 | -27.33 | 45.93 | -77.42 |
| -47.96 | -28.62 | 114.46 | -41.26 | -28.66 | -38.27 | 132.28 | -87.00 |
| -88.38 | -60.16 | -70.32 | -84.83 | -37.93 | 18.20 | -55.00 | 53.86 |
| -14.13 | 32.56 | -88.11 | -27.69 | -18.38 | 73.50 | 75.06 | -12.13 |
| -0.51 | -36.58 | -20.09 | -24.45 | -41.67 | -91.99 | -77.16 | 172.18 |
| -16.80 | -47.94 | 24.53 | -56.50 | -0.37 | 21.02 | 60.54 | 133.41 |
| 222.32 | 79.43 | -17.95 | 93.50 | -46.00 | 122.80 | 58.18 | -19.14 |

(d)

**FIGURE 7.17**    Other suitable solutions obtained by GA-REC.

After generating $G_{Ini}$, the first population is generated by randomly changing limited genes in $G_{Ini}$. Therefore, most bits in the population will become zeros. Figure 7.18 shows the result, which not only corrects the errors but also minimizes the differences between original and translated images. In Figure 7.18, (a) is the original host image, an $8 \times 8$ gray-level image, in the spatial domain, and (b) is (a) transformed by the DCT; (c) and (e) are the translated results of (b) via rounding and GAs, respectively; (d) and (f) are (c) and (e) transformed by the DCT, respectively; (g) and (h) are the extracted data from (d) and (f), respectively. By comparing (a) and (c), one change, from 41 to 40, causes an error, so we cannot extract the embedded data exactly. Once the best solution is found, as shown in (e), an additional change, from 17 to 16, corrects the errors.

Figures 7.19 and 7.20 are other examples of correcting the embedded watermarks 0011 and 1011, respectively. Figures 7.19a and 7.20a are the original images. Figures 7.19b and 7.20b are the images translated by using rounding. Figures 7.19c and 7.20c are extracted from the coefficients of the transformed images Figures 7.19b and 7.20b, respectively. Figures 7.19d and 7.20d are the images translated using a GA. Figures 7.19e and 7.20e are extracted from the coefficients of the transformed images Figures 7.19d and 7.20d, respectively.

In Figure 7.19, because there is no difference between (a) and (b), we know the embedded watermark has been changed after translating real numbers into integers. In (b) and (d), the only difference (51 and 52) corrects the two errors of embedded data from 0000 to 0011.

We can see the same result in Figure 7.20. In (a) and (b), the embedded watermark is changed by the translating procedure from real numbers to integers. Therefore, the extracted data are the same as shown in Figures 7.19c and 7.20c. In Figures 7.20b

| 1 | 8 | 219 | 51 | 69 | 171 | 81 | 41 |
|---|---|-----|----|----|-----|----|----|
| 94 | 108 | 20 | 121 | 17 | 214 | 15 | 74 |
| 233 | 93 | 197 | 83 | 177 | 215 | 183 | 78 |
| 41 | 84 | 118 | 62 | 210 | 71 | 122 | 38 |
| 222 | 73 | 197 | 248 | 125 | 226 | 210 | 5 |
| 35 | 36 | 127 | 5 | 151 | 2 | 197 | 165 |
| 196 | 180 | 142 | 52 | 173 | 151 | 243 | 164 |
| 254 | 62 | 172 | 75 | 21 | 196 | 126 | 224 |

(a) Original cover image

| 970.50 | −42.37 | −4.99 | 94.09 | −94.25 | 82.58 | 115.99 | 96.96 |
|--------|--------|-------|-------|--------|-------|--------|-------|
| −144.74 | 30.63 | −165.94 | 22.53 | −55.09 | −26.76 | 45.39 | −76.50 |
| −46.77 | −28.71 | 113.62 | −40.93 | −28.33 | −39.12 | 131.28 | −87.92 |
| −88.67 | −60.13 | −70.12 | −84.05 | −38.84 | 18.38 | −54.63 | 53.37 |
| −14.75 | 32.48 | −88.16 | −27.56 | −18.00 | 72.99 | 76.57 | −12.66 |
| −1.06 | −37.05 | −19.76 | −24.91 | −41.49 | −91.99 | −76.61 | 171.35 |
| −16.89 | −47.45 | 24.28 | −56.94 | −0.44 | 20.51 | 59.88 | 133.33 |
| 222.41 | 79.21 | −18.53 | 92.78 | −46.48 | 123.71 | 58.15 | −18.58 |

(b) Transformed image from (a)

| 1 | 8 | 219 | 51 | 69 | 171 | 21 | 40 |
|---|---|-----|----|----|-----|----|----|
| 94 | 108 | 20 | 121 | 17 | 214 | 15 | 74 |
| 233 | 93 | 197 | 23 | 177 | 215 | 183 | 78 |
| 41 | 84 | 118 | 62 | 210 | 71 | 122 | 38 |
| 222 | 73 | 197 | 248 | 125 | 226 | 210 | 5 |
| 35 | 36 | 127 | 5 | 151 | 2 | 197 | 165 |
| 196 | 180 | 142 | 52 | 173 | 151 | 243 | 164 |
| 254 | 62 | 172 | 75 | 21 | 196 | 126 | 224 |

(c) Translating result by round

| 970.38 | −42.20 | −5.16 | 94.24 | −94.38 | 82.68 | 115.92 | 96.99 |
|--------|--------|-------|-------|--------|-------|--------|-------|
| −144.91 | 30.87 | −166.17 | 22.73 | −55.26 | −26.62 | 45.30 | −76.46 |
| −46.94 | −28.49 | 113.41 | −40.74 | −28.49 | −38.99 | 131.19 | −87.88 |
| −88.82 | −59.93 | −70.31 | −83.88 | −38.99 | 18.49 | −54.71 | 53.41 |
| −14.88 | 32.65 | −88.32 | −27.41 | −18.13 | 73.09 | 76.50 | −12.62 |
| −1.16 | −36.91 | −19.89 | −24.79 | −41.58 | −91.91 | −76.66 | 171.37 |
| −16.95 | −47.36 | 24.19 | −56.86 | −0.51 | 20.56 | 59.84 | 133.35 |
| 222.37 | 79.26 | −18.58 | 92.82 | −46.51 | 123.73 | 58.13 | −18.57 |

(d) Transformed image from (c)

| 1 | 8 | 219 | 51 | 69 | 170 | 81 | 40 |
|---|---|-----|----|----|-----|----|----|
| 94 | 108 | 20 | 121 | 16 | 214 | 15 | 74 |
| 233 | 93 | 197 | 83 | 177 | 215 | 183 | 78 |
| 41 | 84 | 118 | 62 | 210 | 71 | 122 | 38 |
| 222 | 73 | 197 | 248 | 125 | 226 | 210 | 5 |
| 35 | 36 | 127 | 5 | 151 | 2 | 197 | 165 |
| 196 | 180 | 142 | 52 | 173 | 151 | 243 | 164 |
| 254 | 62 | 172 | 75 | 21 | 196 | 126 | 224 |

(e) Translating result by GAs

| 970.13 | −42.06 | −4.93 | 93.96 | −94.38 | 82.86 | 115.83 | 96.97 |
|--------|--------|-------|-------|--------|-------|--------|-------|
| −145.23 | 31.04 | −165.88 | 22.38 | −55.23 | −26.40 | 45.15 | −76.46 |
| −47.17 | −28.34 | 113.58 | −41.02 | −28.39 | −38.86 | 131.02 | −87.78 |
| −88.93 | −59.83 | −70.27 | −84.06 | −38.81 | 18.49 | −54.92 | 53.63 |
| −14.88 | 32.72 | −88.42 | −27.49 | −17.88 | 72.98 | 76.27 | −12.30 |
| −1.09 | −36.88 | −20.06 | −24.79 | −41.31 | −92.09 | −76.88 | 171.73 |
| −16.86 | −47.35 | 24.02 | −56.83 | −0.28 | 20.39 | 59.67 | 133.66 |
| 222.44 | 79.26 | −18.69 | 92.85 | −46.38 | 123.63 | 58.03 | −18.40 |

(f) Transformed image from (e)

| − | − | 0 | 1 | − | − | − | − |
|---|---|---|---|---|---|---|---|
| − | 1 | − | − | − | − | − | − |
| 1 | − | − | − | − | − | − | − |
| − | − | − | − | − | − | − | − |
| − | − | − | − | − | − | − | − |
| − | − | − | − | − | − | − | − |
| − | − | − | − | − | − | − | − |
| − | − | − | − | − | − | − | − |

(g) The extracted data by GAs

| − | − | 1 | 0 | − | − | − | − |
|---|---|---|---|---|---|---|---|
| − | 0 | − | − | − | − | − | − |
| 0 | − | − | − | − | − | − | − |
| − | − | − | − | − | − | − | − |
| − | − | − | − | − | − | − | − |
| − | − | − | − | − | − | − | − |
| − | − | − | − | − | − | − | − |
| − | − | − | − | − | − | − | − |

(h) The extracted data by round

**FIGURE 7.18**   A more suitable solution obtained by GA-REC.

and 7.20d, the two differences, (118 and 119) and (126 and 125), correct the three errors of embedded data from 0000 to 1011.

## 7.4   APPLICATIONS TO MEDICAL IMAGE WATERMARKING

In recent decades, with the rapid development of biomedical engineering, digital medical images have been becoming increasingly important in hospitals and clinical environments. Concomitantly, traversing between hospitals produces complicated

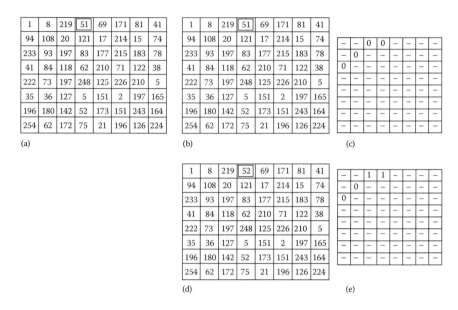

**FIGURE 7.19**    A solution when embedding 0011 by GA-REC.

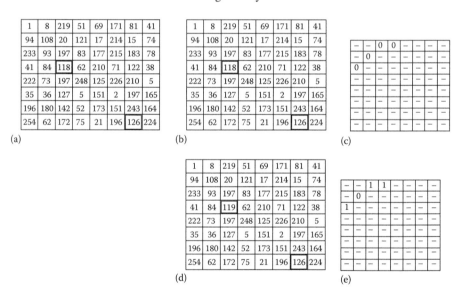

**FIGURE 7.20**    A solution when embedding 1011 by GA-REC.

network protocols, image compression, and security problems. Many techniques have been developed to resolve these problems. For example, the Hospital Information System (HIS) and Picture Archiving and Communication System (PACS) are currently the two primary data communication systems used in hospitals. Although HIS may be slightly different between hospitals, data can be exchanged based on the Health Level Seven (HL7) standard. Similarly, PACS transmits medical images using the Digital Imaging and Communications in Medicine (DICOM) standard. Furthermore, the Institute of Electrical and Electronics Engineers published IEEE 1073 in order to set a standard for measured data and signals from different medical instruments.

Some techniques, such as HIS and PACS, were developed to provide an efficient mechanism for dealing with the outburst of medical images. In order to enhance performance, we need to compress image data for efficient storage and transmission. Lossless compression is adopted for avoiding any distortion when the images are needed for diagnosis. Otherwise, lossy compression can achieve a higher compression rate. Since the quality of medical images is crucial in diagnosis, lossless compression is often used. However, if we can utilize the domain knowledge of specific types of medical images, a higher compression ratio can be achieved without losing any important diagnostic information. In order to obtain higher compression rates, a hybrid compression method [9,10] was developed by compressing the *region of interest* (ROI) with lossless compression and the rest with lossy compression. In order to protect the copyright of medical images, the watermark is embedded surrounding the ROI. Meanwhile, the embedded watermark is preprocessed by *set partitioning in hierarchical trees* (SPIHT) [11] for robustness.

### 7.4.1 OVERVIEW OF THE PROPOSED TECHNIQUE

In order to achieve a higher compression rate without distorting the important data in a medical image, we select the ROI of a medical image and compress it by lossless compression and the rest by lossy compression. For the purpose of protecting medical images and maintaining their integrity, we embed the information watermark (signature image or textual data) and fragile watermark into the frequency domain surrounding the ROI. Meanwhile, the embedded information watermark is preprocessed into a bitstream depending on different types of watermark.

#### 7.4.1.1 Signature Image

##### 7.4.1.1.1 Encoding Procedure

Figure 7.21 shows the encoding procedure when the watermark is a signature image. The host image (medical image) is separated into two parts, ROI and non-ROI. We embed the signature image and fragile watermark into the non-ROI part with GAs. Note that we preprocess the signature image by SPIHT in order to reconstruct it to perceptual satisfaction. Last, we obtain the watermarked image by combining the embedded non-ROI and ROI.

##### 7.4.1.1.2 Decoding Procedure

Figure 7.22 shows the decoding procedure. The non-ROI part is selected from the watermarked image and is transformed by the DCT. Last, we obtain the fragile watermark and a set of bitstreams by extracting data from the specific positions of the coefficients in the frequency domain. We can decide the integrity of the medical image by checking the fragile watermark and obtain the signature image by reconstructing the bitstream using SPIHT.

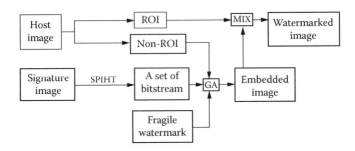

**FIGURE 7.21**    The encoding procedure of embedding a signature image.

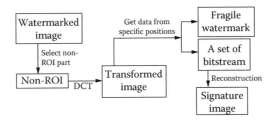

**FIGURE 7.22**    The decoding procedure of embedding a signature image.

### 7.4.1.1.3    SPIHT Compression

SPIHT is a zerotree structure based on the *discrete wavelet transform* (DWT). The first zerotree structure, the *embedded zerotree wavelet* (EZW), was published by Shapiro [12] in 1993. SPIHT, in essence, uses a bit allocation strategy and produces a progressively embedded scalable bitstream.

Figure 7.23 shows an example of SPIHT, where (a) is the original $8 \times 8$ image. We obtain the transformed image (b) with the DWT, and (c) lists the bitstreams generated by applying SPIHT three times. We reconstruct the image by using these bitstreams, and the result is shown in (d).

## 7.4.1.2    Textual Data

### 7.4.1.2.1    Encoding and Decoding Procedures

Figure 7.24 shows the encoding procedure when the watermark is textual data. The procedure is similar to signature image encoding except the encryption of textual data is used. Its decoding procedure is shown in Figure 7.25.

### 7.4.1.2.2    Encryption Scheme in Textual Data

There are many techniques for encrypting textual data. Generally, the main idea is to translate the textual data of plain text into secret codes of cipher text. There are two types of encryption: asymmetric and symmetric.

An easy way to achieve encryption is *bit shifting*. That is, we treat a character as a byte value and shift its bit position. For example, the byte value of a letter "F" is 01000110. We can change it to "d" as 01100100 by shifting the left-most four bits to the right side.

**(a) Original 8*8 image**

| 192 | 192 | 192 | 192 | 192 | 192 | 192 | 192 |
|-----|-----|-----|-----|-----|-----|-----|-----|
| 192 | 16  | 16  | 16  | 192 | 16  | 192 | 192 |
| 192 | 16  | 16  | 192 | 192 | 192 | 192 | 192 |
| 192 | 16  | 16  | 16  | 16  | 192 | 192 | 192 |
| 192 | 16  | 192 | 192 | 192 | 192 | 192 | 192 |
| 192 | 16  | 192 | 192 | 192 | 192 | 192 | 192 |
| 192 | 16  | 16  | 192 | 192 | 192 | 192 | 192 |
| 192 | 192 | 192 | 192 | 192 | 192 | 192 | 192 |

**(b) Transformed image by DWT**

| 139.75 | −24.75 | 11  | −11 | 44  | 0   | 0 | −44 |
|--------|--------|-----|-----|-----|-----|---|-----|
| −2.25  | 2.75   | 11  | −22 | 88  | −88 | 0 | 0   |
| 11     | −33    | 11  | −11 | 88  | −44 | 0 | 0   |
| −33    | −22    | 11  | −22 | 44  | −44 | 0 | 0   |
| 44     | 88     | 88  | 44  | −44 | 0   | 0 | 44  |
| 0      | 0      | 44  | 0   | 0   | 0   | 0 | 0   |
| 0      | −44    | −88 | 0   | 0   | 44  | 0 | 0   |
| −44    | −44    | 0   | 0   | 44  | −44 | 0 | 0   |

**(c) The bitstreams by SPIHT**

110000001
0000100001000001110011100001011001110001100010000000
0000010000101001000001111100010100010101110010110010100010001001001010011110011111110000000000000000

**(c) The reconstructed image**

| 138.25 | 0    | 0     | 0   | 40  | 0     | 0   | 0   | −40 |
|--------|------|-------|-----|-----|-------|-----|-----|-----|
| 0      | 0    | −40   | 0   | 92.5| −92.5 | 0   | 0   | 0   |
| 0      | 0    | 0     | 0   | 92.5| −40   | 0   | 0   | 0   |
| −40    | 92.5 | 92.5  | 40  | 40  | −40   | 0   | 0   | 40  |
| 40     | 0    | 0     | 0   | −40 | 0     | 0   | 0   | 0   |
| 0      | 0    | 0     | 0   | 0   | 0     | 0   | 0   | 0   |
| 0      | −40  | −92.5 | 0   | 0   | 40    | 40  | 0   | 0   |
| −40    | −40  | 0     | 0   | 40  | −40   | 0   | 0   | 0   |

**FIGURE 7.23**　An example of SPIHT.

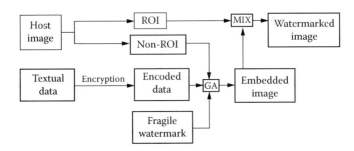

**FIGURE 7.24**    The encoding procedure of embedding textual data.

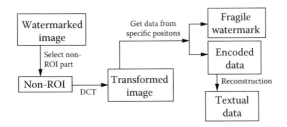

**FIGURE 7.25**    The decoding procedure of embedding textual data.

An encryption method using a logarithm of ASCII codes was developed by Rajendra Acharya [13]. The encryption algorithm can be mathematically stated as

$$T_e = (\log(T_o * 2) * 100) - 300, \tag{7.6}$$

where $T_e$ denotes the encrypted text and $T_o$ denotes the ASCII code of the original text. The decrypted text can be obtained by

$$T_o = \exp\{(T_e + 300)/100 - \log 2\}. \tag{7.7}$$

Note that the encrypted information ($T_e$) is stored as an integer.

### 7.4.1.3    Watermarking Algorithm for Medical Images

Let $H$ be the original host image with size $K \times K$, which is separated into $H^{ROI}$ (ROI) and $H^{NROI}$ (non-ROI) with sizes $N \times M$ and $K \times K - N \times M$, respectively. Let $S$ and $T$ denote a signature image with size $W \times W$ and the textual data, respectively. $S^B$ is a bitstream obtained by compressing $S$ with SPIHT compression or encoding $T_o$ with the encryption technique. $W^F$ is the fragile watermark. $H^{WROI}$ is obtained by adjusting the pixel values of $H^{NROI}$, from which we can extract $S^B$ and $W^F$ from specific positions of the coefficients of the frequency domain in $H^{WROI}$. $H^{Final}$ is the final image given by combining $H^{ROI}$ and $H^{WROI}$.

*Algorithm*:

1. Separate the host image, $H$, into $H^{ROI}$ and $H^{NROI}$.

$$H = \{h(i, j), 0 \le i, j < K\},$$

where $h(i,j) \in \{0, 1, 2,\ldots 2^{L-1}\}$ and $L$ is the number of bits used in the gray level of pixels.

$$H^{ROI} = \{h^{ROI}(i, j),\ 0 \le i < N,\ 0 \le j < M\},$$

where $h^{ROI}(i,j) \in \{0, 1, 2,\ldots 2^{L}-1\}$.

$$H^{NROI} = \{h^{NROI}(i, j)\},$$

where $h^{NROI}(i,j) \in \{0, 1, 2,\ldots 2^{L}-1\}$.

2. *For the signature image*:
Transform $S$ to $S^B$ by SPIHT.

$$S = \{s(i, j),\ 0 \le i,\ j < W\},$$

where $s(i,j) \in \{0, 1, 2,\ldots 2^{L}-1\}$.
$S^B$ is the bitstream consisting of bits 0 and 1.

*For textual data*:
Encode $T$ to $S^B$ with the encryption technique.

3. Adjust $H^{NROI}$ using GAs to obtain $H^{WROI}$, from which we can extract $S^B$ and $W^F$ from the coefficients of its frequency domain.
4. Combine $H^{ROI}$ and $H^{WROI}$ to obtain the final image $H^{Final}$.

Note that we can embed not only the information image but also the fragile watermark into the host image. If there is more than one regular ROI, we can record the following information for each ROI: the top-left corner coordinates, the width, and the height. The watermark image can be placed in a group of $8 \times 8$ blocks that are extracted from the outer layer of each ROI. For an irregular polygon or circular ROI, we need to record the starting and ending positions of each row in the ROI in order to extract the data from the outer layer of its boundary.

## 7.4.2 IMPROVED SCHEME BASED ON GENETIC ALGORITHMS

In this section, we present an improved scheme for embedding watermarks into the frequency domain of a host image. The new scheme not only reduces the cost of obtaining the solution but also offers more applications in watermarking. The main idea is to adjust the pixels in the host image based on GAs, and make sure the extracted data from the specific positions in the frequency domain of the host image are the same as the watermark. Figures 7.26 and 7.27 show the encoding and decoding procedures of the improved scheme, respectively.

The improved algorithm for embedding watermarks based on GAs is presented in the following. Its flowchart is shown in Figure 7.28.

**FIGURE 7.26**   The encoding procedure of the improved scheme.

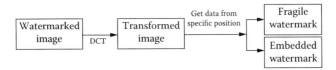

**FIGURE 7.27**   The decoding procedure of the improved scheme.

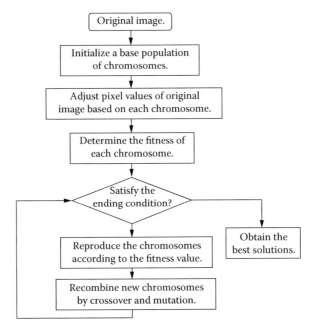

**FIGURE 7.28**   Basic steps to finding the best solutions with improved GAs.

*Algorithm*:

1. Define the fitness function, number of genes, size of population, crossover rate, and mutation rate.
2. Create the first generation by random selection.
3. Adjust the pixel values based on each chromosome of the generation.
4. Evaluate the fitness value for each corresponding chromosome.
5. Obtain the best chromosomes.
6. Recombine new chromosomes by using crossover.
7. Recombine new chromosomes by using mutation.
8. Repeat steps 3 to 8 until a predefined condition is satisfied or a constant number of iterations is reached.

Figure 7.29 shows an example of the improved GAs; (a) and (b) are the original image and the signature data, respectively. Using the embedded order shown in (c), we obtain (d) by separating the signature data into these 12 parts. Figure 7.29e is the fragile watermark, which is defined by the user. Our purpose is to adjust the pixel values of Figure 7.29a in order to obtain its frequency domain, from which we can extract the signature data and fragile watermarks from specific positions. Here, the signature data and the fragile watermark are extracted from the positions of bits 3, 4, 5, and bit 1, respectively. Figure 7.29f shows the result of the adjusted image. We can extract the signature data and fragile watermark from the coefficients of the frequency domain of the watermarked image, as shown in Figure 7.29g. Table 7.2 shows the watermark extracted from the specific positions of the coefficients of the frequency domain of the original image.

### 7.4.3 EXPERIMENTAL RESULTS

Figure 7.30 shows an example of embedding a signature image into an MRI brain image. Figures 7.30a,b show the original medical image with size $230 \times 230$ and the signature image with size $64 \times 64$, respectively. Figure 7.30c is the image transformed by the DWT. The ROI part is marked as a rectangle with size $91 \times 112$, as shown in (d). We encode (c) by SPIHT into a set of bitstreams, which are embedded around the ROI. In (e), the area between two rectangles, $91 \times 112$ and $117 \times 138$, is the clipped watermarked area. The signature image is extracted and the reconstructed result is shown in (f). Table 7.3 shows the PSNR of the original and reconstructed signature images, and of the nonwatermarked and watermarked parts. The error measures, PSNR, are defined as follows:

$$
PSNR = 10 * \log_{10} \left( \frac{\sum_{i=1}^{N} \sum_{j=1}^{N} \left[ h^{GA}(i,j) \right]^2}{\sum_{i=1}^{N} \sum_{j=1}^{N} \left[ h(i,j) - h^{GA}(i,j) \right]^2} \right). \tag{7.8}
$$

Figure 7.31 shows an example of embedding textual data into a computer tomography brain image. Figures 7.31a and 7.31b show the original medical image with size $260 \times 260$ and the textual data, respectively. Figure 7.31c is the encrypted data. The ROI part is marked as a rectangle with size $179 \times 109$, as shown in Figure 7.31d. We obtain a set of bitstreams by shifting the right-most four bits to the left side. In Figure 7.31e, the area between two rectangles is the clipped watermarked area. Note that the original and extracted textual data are exactly the same.

Indeed, GAs consume more time. Our techniques are focused on the following two goals: first, the embedded watermarks should be as robust as possible and can be used to detect any unauthorized modification; second, the compression rate of an image should be as high as possible. Furthermore, the embedded watermarks will be disturbed when we convert real numbers into integers. GAs are adopted to achieve these goals.

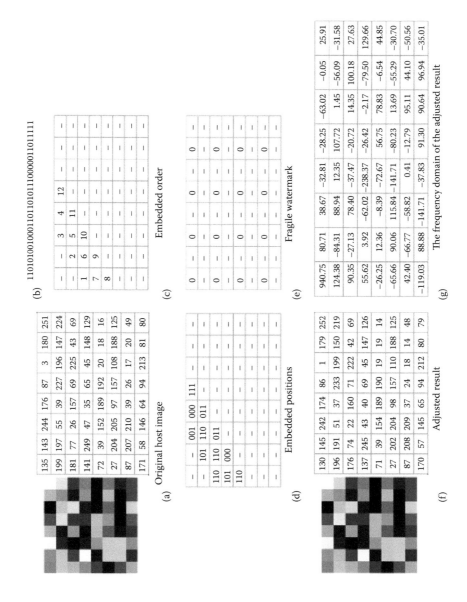

FIGURE 7.29   An example of the improved GAs.

**TABLE 7.2**

**Watermark Extracted from Specific Positions of the Binary Form**

| Decimal | Binary | Decimal | Binary | Decimal | Binary |
|---|---|---|---|---|---|
| 90 | 01011010 | 88 | 01011000 | 3 | 00000011 |
| 84 | 01010100 | 27 | 00011011 | 78 | 01001110 |
| 38 | 00100110 | 55 | 00110111 | 12 | 00001100 |
| 32 | 00100000 | 26 | 00011010 | 28 | 00011100 |

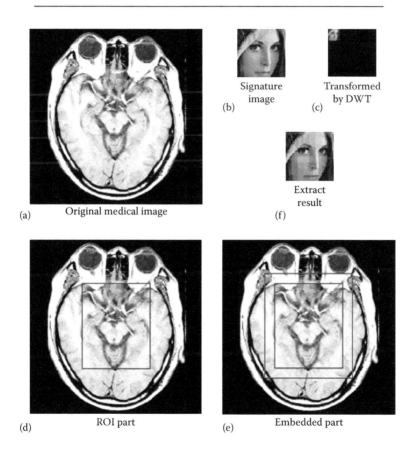

(a) Original medical image

(b) Signature image

(c) Transformed by DWT

(f) Extract result

(d) ROI part

(e) Embedded part

**FIGURE 7.30**   An example of embedding a signature image.

In general, the computational time is closely related to the amount of the required embedded data. That is, the more embedded data we have, the more computational time it takes. For example, it takes about 4 minutes on a Pentium III PC with 600 MHz to obtain the results in Figure 7.29, since there are 36 digits in a bitstream and 16 digits in the fragile watermark. Note that in the literature, usually only four digits of a bitstream and none of a fragile watermark are used.

**TABLE 7.3**
**PSNRs of Signature Image and of Medical Image**

| | Original and Reconstructed Signature Images | Nonwatermarked and Watermarked Medical Images |
|---|---|---|
| PSNR | 24.08 | 38.28 |

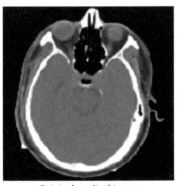

(a) Original medical image

| THE PATIENT INFORMATION |
|---|
| Patient Ref. No:AX8865098 |
| Name of the doctor:Dr.Wu |
| Name of the doctor:Mr.Cheng |
| Age: 48 years |
| Address: 22 Midland Ave. |
| Case History: |
| Date of admission: 18.05.2001 |
| Results: T wave inversion |
| Diagnosis: Suspected MI |

(b) Textual data

| E      E  胡    d??E   ? |
|---|
|    G   澗 %Vf?噭?? S ? |
| ?焓 鎵 G   F?G?A 謘W |
| ?焓 鎵 G   F?G??'? 癸 |
|    VVz ?     '7 |
|    FF'V7?# ?F?酒  gV? |
| 4 7V ?7G? |
| D GV 鎵  F?77  璉    S? |
| %V7w謳7B  w gV   gV'7  ? |
| D?v酘7??W7 V6GVF ? |

(c) Encrypted data

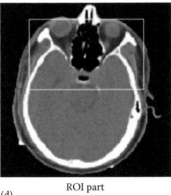

(d) ROI part

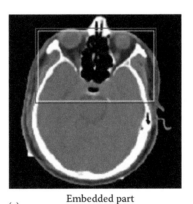

(e) Embedded part

**FIGURE 7.31** The example of embedding textual data.

## 7.5 AUTHENTICATION OF JPEG IMAGES BASED ON GENETIC ALGORITHMS

Authentication of digital images can be carried out in two ways: *digital signing* and *digital watermarking*. A digital signature, such as a hash value, which is a cryptographic technique, has two drawbacks in authenticating images. First, in order to check the authenticity, the verifier must know the digital signature beforehand, so the sender has to send it through a separate secure communication channel. This

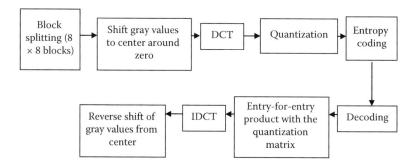

**FIGURE 7.32** The JPEG procedure.

generates an overhead on bandwidth. Second, although digital signatures authenticate images, they fail to locate which parts of the images were altered. One way of achieving tamper localization is dividing the image into blocks and sending the digital signature associated with each block to the receiver. However, this increases the overhead on bandwidth.

Figure 7.32 shows the commonly used JPEG procedure of compressing an image. It involves splitting the input image into blocks of size $8 \times 8$ and using the DCT, quantization, entropy coding, and so on. [14]. As previously mentioned, the existing fragile and semifragile watermarking methods [15–17] need the receiver to know the quantization table beforehand to authenticate the image. As different JPEG quality factors use different quantization tables, the receiver has to maintain a lot of quantization tables. We intend to overcome this problem to allow the receiver to authenticate the image irrespective of the quality factor and the quantization table used in creating the watermarked image. We propose a GA-based method that can adjust the image such that the modified image after JPEG compression contains authentication information in the DCT coefficients of each block. Furthermore, the generation of authentication information guarantees uniqueness with respect to each block (depending on the block's position) and each image, thwarting *copy paste* (CP) and *vector quantization* (VQ) attacks.

Section 7.5.1 presents the GA-based watermark-embedding method. Section 7.5.2 describes the authentication procedure. Section 7.5.3 shows the experimental results. Finally, conclusions are drawn in Section 7.5.4.

## 7.5.1 GA-BASED WATERMARK-EMBEDDING METHOD

### 7.5.1.1 Overall Watermark-Embedding Procedure

A block diagram of the GA-based watermark-embedding procedure is shown in Figure 7.33. Each step is briefly described in the following series of steps and will be explained in more detail in later sections. Let $A$ be an original grayscale image of size $N \times N$, where $N$ is a multiple of 8. Let the *quality factor* of JPEG compression be denoted as QF.

> $Step_e$ *1*. As JPEG uses blocks of size $8 \times 8$, we divide the image $A$ into a set of $8 \times 8$ blocks, named $\{b(i,j) \mid i,j = 1, 2,..., N/8\}$, where $i$ and $j$ denote the row and column numbers of the block, respectively. Mark each block $b(i,j)$

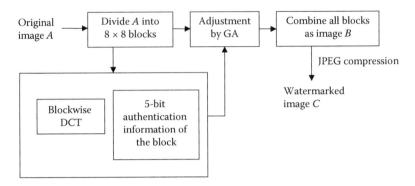

**FIGURE 7.33**   The overall watermark-embedding procedure.

with a number $X$ by traversing the blocks in the image from top to bottom and left to right as

$$X(i, j) = (i-1) \times (N/8) + (j-1). \qquad (7.9)$$

$Step_e$ 2. Apply a chaotic map [4] to transform the block number $X$ into a new mapping block number $X$ as

$$X'(i, j) = \{(k \times X(i, j)) \bmod M\} + 1, \qquad (7.10)$$

where:
  $M = N/8 \times N/8$ is the total number of blocks in the image
  $X, X' \in [0, M-1]$
  $k \in [2, M-1]$ (a secret key) is a prime number

Note that if $k$ is not a prime number, the one-to-one mapping cannot be achieved.

$Step_e$ 3. For each block $b(i,j)$, we generate the authentication information which will be described in Section 7.5.1.2.

$Step_e$ 4. For each block $b(i,j)$, we adjust its values using GA, so the modified block $b'(i,j)$ after JPEG compression contains the authentication information generated in step 3 in the LSB of the integer part of the five upper-left DCT coefficients, as shown in Figure 7.34.

$Step_e$ 5. Combine all modified blocks $b'(i,j)$ to generate modified image $B$.

$Step_e$ 6. Compress the modified image $B$ using JPEG compression with the QF to obtain the watermarked image $C$.

### 7.5.1.2  Authentication Information Generation

In order to thwart CP and VQ attacks, the authentication information needs to be unique per block and per image. Figure 7.35 illustrates the authentication information generation. The chaotic-map block number $X'(i,j)$ computed in Section 7.5.1.1 is unique to each block. The 11-bit clock information is obtained by using the bitwise

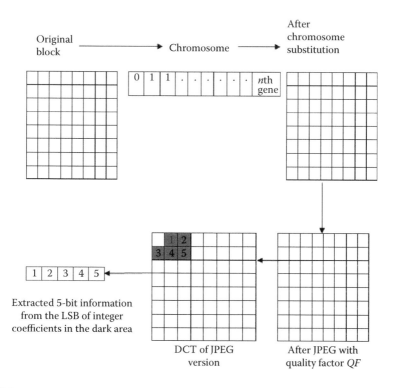

**FIGURE 7.34** The GA procedure to adjust the block.

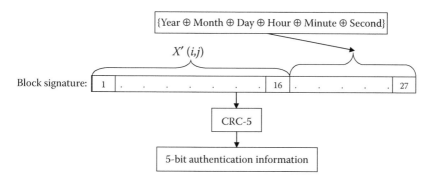

**FIGURE 7.35** The generation of authentication information.

*exclusive or* (bit-XOR $\oplus$) operation with six elements of image capture identity (recorded in the image header): year, month, day, hour, minute, and second. We attach the 11-bit information to the end of the 16-bit stream of $X'(i,j)$ to generate a 27-bit block signature. Finally, the 5-bit authentication information is computed by subjecting the 27-bit block signature to CRC-5. The secret key $k$ of the chaotic map is shared with the receiver using the public-key cryptographic algorithm [18].

In the computation of CRC [18], a $k$-bit string is represented by the coefficients of a $k-1$ degree polynomial (i.e., the coefficients of $x^0$ to $x^{k-1}$). The MSB of the

$k$-bit string becomes the coefficient of $x^{k-1}$, the second MSB becomes the coefficient of $x^{k-2}$, and so on, and the LSB becomes the coefficient of $x^0$. For example, the polynomial representation of the bit string 1101 is $x^3 + x^2 + 1$. To compute the CRC-$r$ authentication information of a bit string, we pad $r$ zeros to the end of the bit string. Then we divide the resulting bit string with an $r$-degree polynomial. We subtract the remainder from the bit string padded with zeros. The resultant bit string after subtraction is the authentication information. The subtraction is done in modulo-2 arithmetic. To check if a CRC-$r$-generated bit string is modified or not, we divide the bit string with the $r$-degree polynomial used in the generation stage. If there is a nonzero remainder after division, it means that the bit string is modified.

### 7.5.1.3   Adjustment by GA

GA is a randomized, parallel, and global search approach based on the mechanics of natural selection and natural genetics to find solutions to problems. Generally, a GA starts with a randomly selected population of genes in the first generation. Each individual chromosome in the population corresponds to a solution in the problem domain. An objective (or fitness function) is used to evaluate the quality of each chromosome. High-quality chromosomes will survive and form the population of the next generation. By using the reproduction, crossover, and mutation operations, a new generation is recombined to find the best solution. This process will repeat until a prespecified condition is satisfied or a constant number of iterations are reached.

We apply a GA to adjust the pixel values in each block $b(i,j)$, such that the adjusted block $b'(i,j)$, after compression with the QF, contains the authentication information (as shown in Section 7.5.1.2) in the LSB of the integer part of the five DCT coefficients, as indicated in Figure 7.34. This algorithm is divided into three steps: the substitution of chromosomes in the original block, chromosome evaluation, and a GA. We present the GA adjustment process with two different approaches. The two approaches use the details in Sections 7.5.1.3.1 and 7.5.1.3.2 in the evaluation of each chromosome to find a suitable adjusted block.

### 7.5.1.3.1   *Substitution of Chromosomes in the Original Block*

The idea is to modify the $p$ LSBs of each pixel in the block, where $p$ is equal to the chromosome length divided by 64. For example, if the chromosome length is 64, then only one LSB is modified in each pixel. We traverse the pixels from left to right and top to bottom, and substitute the first gene in the LSB of the first pixel, the second gene in the LSB of the second pixel, and so on. As another example, if the chromosome length is $64p$, then the first $p$ genes are used to replace the $p$ LSBs of the first pixel, the next $p$ genes are used to replace the $p$ LSBs of the second pixel, and so on.

### 7.5.1.3.2   *Objective Function for Chromosome Evaluation*

The objective function is the bit difference between the five extracted LSBs from the integer part of the DCT coefficients (as shown in Figure 7.34) and the 5-bit authentication information (as described in Section 7.5.1.2). We intend to minimize the difference, with zero being treated as the best solution. The objective function can be expressed as

$$\text{Objective\_function} = \sum_{i=1}^{5} |a(i) - r(i)|, \qquad (7.11)$$

where $a$ is the 5-bit authentic information and $r$ is the retrieved LSBs from the integer part of the five DCT coefficients.

### 7.5.1.3.3 Genetic Algorithm to Adjust the Block

We present two types of block adjustment: *fixed-length chromosome* and *variable-length chromosome*. It is a trade-off between speed and watermarked image quality. The fixed-length chromosome algorithm runs faster but produces a lower-quality watermarked image compared with the variable-length chromosome algorithm.

*7.5.1.3.3.1 Fixed-Length Chromosome*  In the fixed-length chromosome algorithm, the chromosome length is fixed and determined by the QF of JPEG compression. The chromosome length designed for the corresponding QF range is listed in Table 7.1. It aims at adjusting the block with acceptable image quality under the reduced time consumption of the GA adjustment process. The GA adjustment process by the fixed-length chromosome algorithm is presented as follows (Table 7.4):

*Step$_f$ 1.* Define the chromosome length, population size, crossover rate, replacement factor, and mutation rate. The initial population is randomly assigned with 0s and 1s.

*Step$_f$ 2.* Substitute each chromosome in the original block as explained in Section 7.5.1.3.1 and evaluate the objective function for each corresponding chromosome with Equation 7.11.

*Step$_f$ 3.* Apply reproduction, crossover, and mutation operators to create the next generation of chromosomes.

*Step$_f$ 4.* Repeat steps 2 and 3 until the objective function value is equal to zero.

**TABLE 7.4**

**Chromosome Lengths Associated with Different QF Values**

| Quality Factor | Chromosome Length |
| --- | --- |
| QF > 80 | 192 |
| $65 \leq QF \leq 80$ | 256 |
| $55 \leq QF \leq 65$ | 320 |
| $45 \leq QF \leq 55$ | 384 |
| QF < 45 | 448 |

*Step$_f$ 5*. Substitute the final chromosome in the original block to obtain the modified block.

*7.5.1.3.3.2  Variable-Length Chromosome*    In the variable-length chromosome algorithm, to improve the watermarked image quality we start the adjustment process with a chromosome size of 64. If a solution cannot be obtained before the termination condition is satisfied, we increase the chromosome size by 64 and execute the adjustment process again. This process is repeated until the termination condition is satisfied. The final chromosome size is selected based on the QF of the JPEG compression in Table 7.1. The termination condition for the final chromosome size is to obtain an objective function value of zero and that of the intermediate chromosome sizes is a generation limit of 100. The GA adjustment process by the variable-length chromosome algorithm with a QF of 65 is presented as follows and its flowchart is shown in Figure 7.36.

*Step$_v$ 1*. Define the objective function, number of genes ($n = 64$), population size, crossover rate, replacement factor, and mutation rate. The initial population is randomly assigned with 0s and 1s.

*Step$_v$ 2*. If $n < 256$, the termination condition is the maximum of 100 generations; otherwise, the termination condition is the objective function value being zero.

*Step$_v$ 3*. Substitute each chromosome in the original block as explained in Section 7.5.1.3.1 and evaluate the objective function for each corresponding chromosome with Equation 7.11.

*Step$_v$ 4*. While the termination condition is not satisfied, apply reproduction, crossover, and mutation operators to create the next generation of chromosomes, and compute the objective function for each corresponding chromosome.

*Step$_v$ 5*. If the objective function value of the population is zero, substitute the chromosome in the original block to obtain the modified block; otherwise, $n = n + 64$ and go to step 2.

## 7.5.2  AUTHENTICATION

Authentication is the procedure that validates the authenticity of the received image. The main objective of authentication is to detect tampered blocks with a low *false acceptance ratio* (FAR). Our authentication algorithm is presented as follows:

*Step$_a$ 1*. Let $D$ be a testing image, which is divided into blocks of size $8 \times 8$. Let the group of blocks be denoted as *attack_blocks*.

*Step$_a$ 2*. Retrieve the secret key $k$, the polynomial used in CRC, and the clock values.

*Step$_a$ 3*. Compute the block numbers and the corresponding mapping block number, as in Section 7.5.1.1.

*Step$_a$ 4*. (Level-1 detection) For each block *attack_blocks(i,j)* $\in D$, where $1 \le i,j \le N/8$, compute the authentication information, as in Section 7.5.1.2,

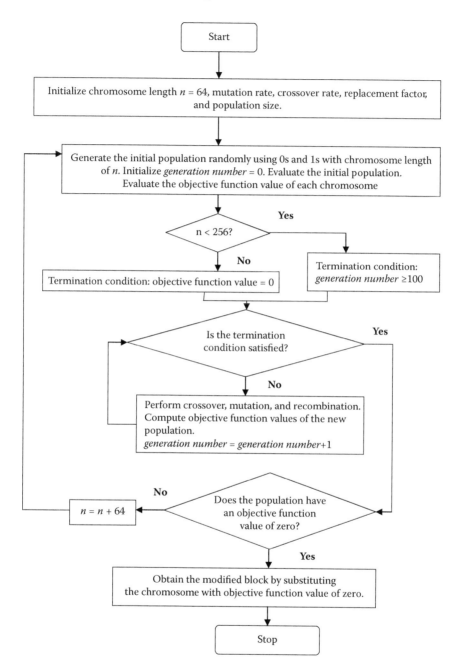

**FIGURE 7.36** The variable-length chromosome procedure.

and denote it as *auth*. Compute the DCT of *attack_blocks(i,j)*. Retrieve the 5-bit information from the LSB of the integer part of the five DCT coefficients, as shown in Figure 7.34, and denote it as *retv*. Follow the same order of DCT coefficients used in the embedding stage while retrieving. If $\sum_{k=1}^{5} |auth(k) - retv(k)| \neq 0$, mark *attack_blocks(i,j)* as invalid; otherwise, mark *attack_blocks(i,j)* as valid.

$Step_a$ 5. (Level-2 detection) This level is used to reduce the FAR of invalid blocks. For each valid block after Level-1 detection, if it has five or more invalid blocks in its $3 \times 3$ neighborhood, mark the block as invalid; otherwise, mark the block as valid.

The proposed authentication algorithm not only can detect the tampered blocks, but it can also reduce the burden on the verifier as it does not require the DCT coefficients to be quantized in the authentication process. Furthermore, this two-level detection strategy can reduce the FAR of invalid blocks.

### 7.5.3  EXPERIMENTAL RESULTS

To evaluate experimental results, we use the PSNR to measure the quality of the watermarked image. Let the image size be $m \times n$. The PSNR in decibels (dB) of an image $A$ with respect to an image $B$ is defined by

$$PSNR = 20 \log_{10} \left( \frac{255}{\sqrt{MSE}} \right), \tag{7.12}$$

where *MSE* is the mean square error defined by

$$MSE = \frac{1}{mn} \sum_{i=0}^{m-1} \sum_{j=0}^{n-1} [A(i,j) - B(i,j)]^2. \tag{7.13}$$

Note that the higher the PSNR is, the less distortion there is to the host image and the retrieved one.

We use the FAR as a metric to evaluate the performance of the authentication system. The FAR is defined as the ratio of the total number of undetected blocks to the total number of attacked blocks. Its value varies is between 0 and 1. The lesser the FAR value, the better the performance of the authentication system.

$$FAR = \frac{\text{Total number of undetected blocks}}{\text{Total number of attacked blocks}}. \tag{7.14}$$

Figure 7.37 shows the *Lena* image before and after applying the watermarking algorithms with fixed-length and variable-length chromosomes, where QF = 75, secret key $k = 13$, and clock = 2008/11/27 12:45:40. As expected, the variable-length chromosome algorithm provides better watermarked image quality than the fixed-length one. More of the PSNR values of variable-length chromosomes on the *Barbara*, *Baboon*, and *Cameraman* images are listed in Table 7.5.

**FIGURE 7.37** (a) The original *Lena* image, (b) the image watermarked by a fixed-length chromosome algorithm (PSNR = 36.67), (c) the image watermarked by a variable-length chromosome algorithm (PSNR = 45.12).

**TABLE 7.5**
**PSNR Values of Four Watermarked Images**

|  | Lena | Barbara | Baboon | Cameraman |
|---|---|---|---|---|
| PSNR | 45.12 | 35.34 | 45.12 | 38.13 |

We conduct CP, cropping, and VQ attacks on the watermarked image. In a CP attack [19], a portion of a given image is copied and then used to cover some other objects in the given image. If the splicing is perfectly performed, human perception will not detect that identical (or virtually identical) regions exist in the image. The ideal regions for a CP attack are textured areas with irregular patterns, such as grass. Because the copied areas will likely blend with the background, it will be very difficult for the human eye to detect any suspicious artifacts. Another fact which complicates detection is that the copied regions come from the same image. They therefore have similar properties, such as the noise component or color palette. This makes the use of statistical measures to find irregularities in different parts of the image impossible. In the case of blockwise authentication systems, if the system has no blockwise dependency or unique signature that differentiates each block, then this kind of attack is not detected.

Holliman and Memon [20] presented a counterfeiting attack on blockwise independent watermarking schemes, referred to as a VQ or *collage* attack. In such a scheme the attacker generates a forged image by collaging authenticated blocks from different watermarked images. Because of the blockwise nature of the embedding and authentication processes, the forgery is authenticated as valid. Additionally, if the database of watermarked images is huge, the attacker can easily generate a collage that is identical to the unwatermarked image.

Figure 7.38 shows the results of a watermarked car image along with the CP-attacked version and the authentication images after Level-1 and Level-2 detections. The white portion in the authentication image indicates the modified part. Figure 7.39 shows the results of a watermarked oryx image. We also tested the proposed algorithm under the cropping attack using 11 different single-chunk cropping criteria, as shown in Figure 7.40, and their resulting FAR values are listed in Table 7.6. Lastly, Figure 7.41 shows the resulting images under different cropping patterns.

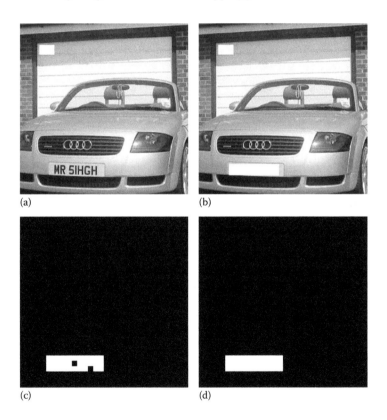

(a)          (b)

(c)          (d)

**FIGURE 7.38**    (a) The image watermarked by the variable-length chromosome algorithm (PSNR = 43.36), (b) the CP-attacked watermarked image with the plate number removed, (c) the authentication image after Level-1 detection, (d) the authentication image after Level-2 detection.

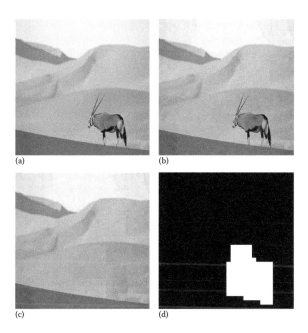

**FIGURE 7.39** (a) The original oryx image, (b) the image watermarked by a variable-length chromosome algorithm (PSNR = 41.14), (c) the CP-attacked image, (d) the authentication image.

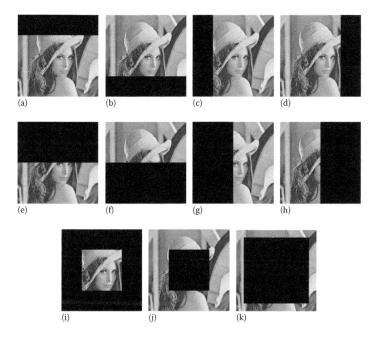

**FIGURE 7.40** Different cropping criteria: (a) top 64, (b) bottom 64, (c) left 64, (d) right 64, (e) top 128, (f) bottom 128, (g) left 128, (h) right 128, (i) outer, (j) center 25%, (k) center 63.7%.

**TABLE 7.6**
**FAR Values under Attack Using Different Cropping Criteria in Figure 7.40**

| Attack Type | a | b | c | d | e | f | g | h | i | j | k |
|---|---|---|---|---|---|---|---|---|---|---|---|
| FAR | 0 | 0 | 0 | 0.0039 | 0 | 0 | 0 | 0 | 0.038 | 0 | 0 |

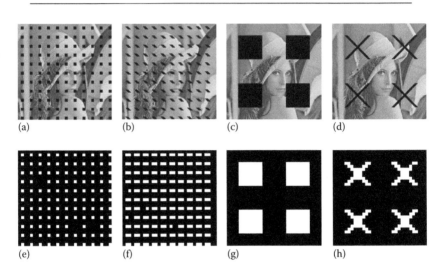

(a)     (b)     (c)     (d)

(e)     (f)     (g)     (h)

**FIGURE 7.41**    (a), (b), (c), (d): Different cropping patterns. (e), (f), (g), (h): Corresponding detection images.

## REFERENCES

1. Rivest, R. L., The MD5 message digest algorithm, RFC 1321, accessed on January 6, 2017, http://www.faqs.org/rfcs/rfc1321.html, 1992.
2. Rivest, R. L., Shamir, A., and Adleman, L., A method for obtaining digital signatures and public-key cryptosystems, *Communications of the ACM*, 21, 120, 1978.
3. Lin, S. D. and Chen, C.-F., A robust DCT-based watermarking for copyright protection, *IEEE Trans. Consumer Electronics*, 46, 415, 2000.
4. Holland, J. H., *Adaptation in Natural and Artificial Systems*, University of Michigan Press, Ann Arbor, 1975.
5. Herrera, F., Lozano, M., and Verdegay, J. L., Applying genetic algorithms in fuzzy optimization problems, *Fuzzy Systems and Artificial Intelligence*, 3, 39, 1994.
6. Ho, S.-Y., Chen, H.-M., and Shu, L.-S., Solving large knowledge base partitioning problems using the intelligent genetic algorithm, in *Proc. Int. Conf. Genetic and Evolutionary Computation*, Orlando, FL, 1999, 1567.
7. Tang, K.-S. et al., Minimal fuzzy memberships and rules using hierarchical genetic algorithms, *IEEE Trans. Industrial Electronics*, 45, 162, 1998.
8. Holland, J. H., *Adaptation in Natural and Artificial System: An Introductory Analysis with Applications to Biology, Control, and Artificial Intelligence*, MIT Press, Cambridge, MA, 1992.

9. Wakatani, A., Digital watermarking for ROI medical images by using compressed signature image, in *Proc. Int. Conf. System Sciences*, Hawaii, 2002.
10. Strom, J. and Cosman, P. C., Medical image compression with lossless regions of interest, *Signal Processing*, 59, 155, 1997.
11. Said, A. and Pearlmana, W. A., A new, fast, and efficient image codec based on set partitioning in hierarchical trees, *IEEE Trans. Circuits and System for Video Technology*, 6, 243, 1997.
12. Shapiro, J., Embedded image coding using zerotrees of wavelet coefficients, *IEEE Trans. Signal Processing*, 41, 3445, 1993.
13. Rajendra Acharya, U. et al., Compact storage of medical image with patient information, *IEEE Trans. Information Technology in Biomedicine*, 5, 320, 2001.
14. Wang, H., Ding, K., and Liao, C., Chaotic watermarking scheme for authentication of JPEG images, in *Proc. Int. Symp. Biometric and Security Technologies*, Islamabad, Pakistan, 2008.
15. Ho, C. K. and Li, C. T., Semi-fragile watermarking scheme for authentication of JPEG images, in *Proc. Int. Conf. Information Technology: Coding and Computing*, Las Vegas, NV, 2004.
16. Lin, C. Y. and Chang, S.-F., Semi-fragile watermarking for authenticating JPEG visual content, in *Proc. SPIE Security and Watermarking of Multimedia Content II*, San Jose, CA, 140, 2000.
17. Li, C. T., Digital fragile watermarking scheme for authentication of JPEG images, in *IEE Proceedings: Vision, Image and Signal Processing*, 151, 460, 2004.
18. Lin, P. L., Huang, P. W., and Peng, A. W., A fragile watermarking scheme for image authentication with localization and recovery, in *Proc. IEEE Sixth Int. Symp. Multimedia Software Engineering*, Miami, FL, 146, 2004.
19. Lee, T. Y. and Lin, S. D., Dual watermark for image tamper detection and recovery, *Pattern Recognition*, 41, 3497, 2008.
20. Holliman, M. and Memon, N., Counterfeiting attacks on oblivious block wise independent invisible watermarking schemes, *IEEE Trans. Image Processing*, 9, 432, 2000.

# 8 Adjusted-Purpose Watermarking

## 8.1 AN ADJUSTED-PURPOSE DIGITAL WATERMARKING TECHNIQUE

Several purpose-oriented watermarking techniques have been developed. Cox et al. [1] proposed spread-spectrum watermarking to embed a watermark into regions with large coefficients in the transformed image. Wong [2] presented a block-based fragile watermarking technique by adopting the Rivest–Shamir–Adleman (RSA) public-key encryption algorithm [3] and message digest 5 (MD5) algorithm [4] for the hashing function. Celik et al. [5] proposed a hierarchical watermarking approach to improve Wong's algorithm. Chen and Lin [6] presented mean quantization watermarking by encoding each bit of the watermark into a set of wavelet coefficients. Note that all the existing watermarking techniques can achieve only one watermarking purpose (robust or fragile) and can be adapted to only one domain (spatial or frequency).

Selecting a suitable watermarking technique is not easy since there are many different kinds of techniques for varying types of watermarks and purposes. In this chapter, a novel *adjusted-purpose* (AP) watermarking technique is developed to integrate different types of techniques. In Section 8.1.1, we present an overview of the technique. A morphological approach to extracting the *pixel-based* (PB) features of an image is presented in Section 8.1.2. In Section 8.1.3, the strategies for adjusting the *variable-sized transform window* (VSTW) and the *quantity factor* (QF) are given. Experimental results are provided in Section 8.1.4. Section 8.1.5 presents a collecting approach to generating VSTW.

### 8.1.1 Overview of Adjusted-Purpose Digital Watermarking

In this section, we present an overview of the AP digital watermarking technique. Before presenting the basic concept, we are concerned with two primary issues:

1. How can we adjust the approach to performing watermarking from the spatial to the frequency domain?
2. How can we adjust the purpose of watermarking from fragile to robust?

In order to achieve these, we develop two parameters: the VSTW and the QF. By adjusting the size of the VSTW, our approach to performing watermarking will use the spatial domain when the VSTW = $1 \times 1$ and the frequency domain for other sizes. On the other hand, by adjusting the value of the QF, the embedded watermarks will become robust if the QF is large and become fragile if the QF is 1. The following are the encoding and decoding procedures of the AP watermarking algorithm.

Let $H$ be a gray-level host image of size $N \times N$ and $W$ be a binary watermark image of size $M \times M$. Let $HB$ be the block-based image obtained by dividing $H$ into non-overlapping blocks of size $n \times n$ and $HB^{DCT}$ be the transformed image by DCT. Let $Q$ be the quantity factor of the same size as $HB$ and $HB^Q$ be the image obtained by dividing $HB^{DCT}$ by $Q$. Let $PB$ be the PB features extracted from $H$, and $FW$ be the final watermark by applying an XOR operator of $PB$ and $W$. Let $\otimes$ denote the operator that substitutes the bits of a watermark for the bits of a host image using *least-significant-bit* (LSB) modification. Note that LSB modification tends to exploit the bit plane, such that the modification does not cause a significantly different visual perception. Let $HB^{WF}$ be the watermarked image, where $HB^Q$ and $FW$ are combined using LSB modification. Let $HB^{WMF}$ be the watermarked image obtained by multiplying $HB^{WF}$ by $Q$. Let $HB^{WS}$ be the watermarked image obtained by converting $HB^{WMF}$ back to the spatial domain by the *inverse discrete cosine transform* (IDCT).

The algorithm for the encoding procedure of our AP watermarking technique is presented in the following, and its flowchart is shown in Figure 8.1.

*Encoding procedure of the AP watermarking algorithm*:

1. Obtain $PB$ from $H$ based on morphological operators.
2. a.  Case 1: Without an additional watermark; set $FW = PB$.
   b.  Case 2: With a watermark $W$; obtain $FW$ by applying the XOR operator of $PB$ and $W$.
3. Determine $n$ and obtain $HB$ by splitting $H$ into non-overlapping subwatermarks of size $n \times n$.
4. Transform $HB$ by DCT to obtain $HB^{DCT}$.
5. Obtain $HB^Q$ by dividing $HB^{DCT}$ by $Q$.
6. Insert $FW$ into the coefficients of $HB^Q$ by LSB modification to obtain $HB^{WF}$.
7. Obtain $HB^{WMF}$ by multiplying $HB^{WF}$ by $Q$.
8. Obtain $HB^{WS}$ by converting $HB^{WMF}$ back to the spatial domain by IDCT.

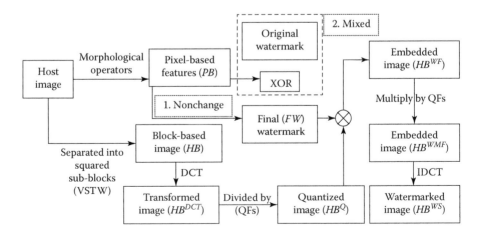

**FIGURE 8.1**   The encoding procedure.

Let $HB^{DWS}$ be the block-based image by dividing $HB^{WS}$ into non-overlapping blocks of size $n \times n$. Let $HB^{DCT}$ be the transformed image of $HB^{DWS}$ by DCT and $HB^{DQ}$ be the image obtained by dividing $HB^{DDCT}$ by $Q$. Let $HB^{LSB}$ be the binary image obtained by extracting the LSB value of each pixel in $HB^{DQ}$. Let $FW^D$ be the extracted watermark by applying an XOR operator of $PB$ and $HB^{DQ}$.

The algorithm for the decoding procedure of our AP watermarking technique is presented in the following, and its flowchart is shown in Figure 8.2.

*Decoding procedure of the AP watermarking algorithm*:

1. Obtain $PB$ from $HB^{WS}$ based on morphological operators.
2. Determine $n$ and obtain $HB^{DWS}$ by splitting $HB^{WS}$ into non-overlapping sub-watermarks of size $n \times n$.
3. Transform $HB^{DWS}$ by DCT to obtain $HB^{DDCT}$.
4. Obtain $HB^{DQ}$ by dividing $HB^{DDCT}$ by $Q$.
5. Obtain $HB^{LSB}$ by extracting the LSB value of each pixel in $HB^{DQ}$.
6. a. Case 1: Without an additional watermark; set $FW^D = HB^{LSB}$.
   b. Case 2: With a watermark $W$; obtain $FW^D$ by applying the XOR operator of $PB$ and $HB^{DQ}$.

### 8.1.2 MORPHOLOGICAL APPROACH TO EXTRACTING PIXEL-BASED FEATURES

Mathematical morphology [7–9], which is based on set theory, can extract object features by choosing a suitable structuring shape as a probe. The analysis is geometric in character and it approaches image processing from the vantage point of human perception. Morphological operators deal with two images. The image being processed is referred to as the *active image*, and the other image, being a kernel, is referred to as the *structuring element*. With the help of various designed structuring shapes, one can simplify image data representation and explore shape characteristics.

In order to extract the PB features of an image, mathematical morphology is adopted. Let $S$ be a grayscale structuring element. Let $H^{Di}$ and $H^{Er}$ be the dilation and the erosion of $H$ by $S$, respectively. Let $PB(i,j)$, $H(i,j)$, $H^{Di}(i,j)$, and $H^{Er}(i,j)$

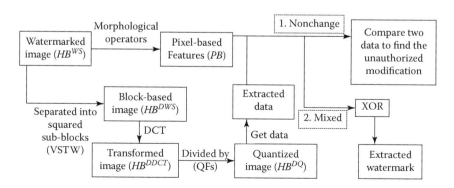

**FIGURE 8.2** The decoding procedure.

respectively denote the grayscale values of $PB$, $H$, $H^{Di}$, and $H^{Er}$ at pixel $(i,j)$. The PB features extraction algorithm is presented in the following.

*PB features extraction algorithm*:

1. Design the grayscale structuring element $S$.
2. Obtain $H^{Di}$ by a dilation of $H$ by $S$.
3. Obtain $H^{Er}$ by an erosion of $H$ by $S$.
4. Obtain the PB features by the following rule:

$$\text{If } (0 \le \frac{H(i,j) - H^{Er}(i,j)}{H^{Di}(i,j) - H^{Er}(i,j)} \le T_1) \text{ or } (T_2 \le \frac{H(i,j) - H^{Er}(i,j)}{H^{Di}(i,j) - H^{Er}(i,j)} \le 1),$$

then $PB(i,j) = 1$; otherwise, $PB(i,j) = 0$.

Note that $T_1$ and $T_2$ are threshold values, and $T_1 < T_2$.

We observe that almost all the PB features of an image will remain the same after compression. Figure 8.3 illustrates the properties of PB features. Figure 8.3a is the original image and Figures 8.3b–d are the compressed images at the compression rates of 80%, 50%, and 20%, respectively. Let us randomly select a pixel $p$ and a $3 \times 3$ neighborhood as shown in the bold box. The value of $p$ (i.e., the central pixel) is originally ranked 5 and is now ranked 4 or 6 after compression, as

| 82 | 76 | 53 | 34 | 33 | 45 | 59 | 86 | 72 | 74 | 124 | 111 |
|----|----|----|----|----|----|----|----|----|----|-----|-----|
| 73 | 64 | 46 | 32 | 54 | 64 | 80 | 93 | 102 | 97 | 84 | 80 |
| 59 | 40 | 34 | 36 | 38 | 50 | 68 | 84 | 96 | 96 | 91 | 86 |
| 53 | 33 | 28 | 31 | 24 | 36 | 53 | 71 | 86 | 90 | 88 | 86 |
| 62 | 55 | 42 | 31 | 30 | 38 | 51 | 64 | 69 | 71 | 70 | 69 |
| 58 | 60 | 55 | 49 | 44 | 48 | 53 | 58 | 57 | 56 | 53 | 51 |
| 44 | 45 | 59 | 73 | 53 | 54 | 55 | 56 | 51 | 49 | 47 | 46 |
| 38 | 38 | 53 | 68 | 50 | 51 | 52 | 54 | 52 | 54 | 56 | 57 |
| 39 | 43 | 43 | 40 | 43 | 44 | 47 | 52 | 58 | 65 | 71 | 75 |
| 55 | 45 | 63 | 58 | 46 | 45 | 44 | 45 | 49 | 55 | 62 | 67 |
| 55 | 32 | 68 | 50 | 36 | 36 | 37 | 42 | 51 | 64 | 75 | 82 |
| 51 | 22 | 67 | 43 | 30 | 30 | 32 | 40 | 54 | 71 | 88 | 98 |

(a)

| 83 | 77 | 50 | 32 | 38 | 45 | 61 | 90 | 70 | 68 | 127 | 117 |
|----|----|----|----|----|----|----|----|----|----|-----|-----|
| 69 | 66 | 48 | 39 | 49 | 57 | 71 | 95 | 102 | 95 | 88 | 82 |
| 53 | 49 | 35 | 31 | 42 | 53 | 67 | 88 | 103 | 92 | 95 | 77 |
| 54 | 42 | 27 | 20 | 26 | 37 | 55 | 72 | 82 | 85 | 86 | 83 |
| 64 | 51 | 39 | 32 | 30 | 37 | 52 | 63 | 66 | 74 | 70 | 72 |
| 61 | 53 | 55 | 56 | 47 | 47 | 55 | 57 | 54 | 60 | 53 | 52 |
| 47 | 45 | 60 | 68 | 55 | 50 | 56 | 53 | 48 | 62 | 30 | 44 |
| 39 | 39 | 59 | 69 | 54 | 49 | 56 | 53 | 54 | 55 | 59 | 59 |
| 44 | 38 | 41 | 46 | 41 | 42 | 51 | 51 | 54 | 71 | 69 | 76 |
| 55 | 45 | 64 | 59 | 46 | 40 | 46 | 44 | 46 | 57 | 56 | 70 |
| 53 | 35 | 69 | 55 | 39 | 32 | 40 | 40 | 56 | 68 | 74 | 89 |
| 50 | 24 | 61 | 41 | 28 | 26 | 37 | 38 | 53 | 72 | 88 | 103 |

(b)

| 82 | 67 | 48 | 35 | 36 | 47 | 61 | 70 | 61 | 94 | 109 | 114 |
|----|----|----|----|----|----|----|----|----|----|-----|-----|
| 70 | 59 | 45 | 38 | 44 | 61 | 80 | 92 | 95 | 101 | 86 | 79 |
| 51 | 43 | 32 | 27 | 34 | 52 | 73 | 87 | 98 | 100 | 85 | 83 |
| 49 | 42 | 32 | 24 | 24 | 35 | 51 | 63 | 82 | 89 | 86 | 96 |
| 62 | 59 | 51 | 41 | 35 | 38 | 48 | 58 | 75 | 73 | 63 | 72 |
| 56 | 59 | 59 | 53 | 46 | 45 | 53 | 61 | 57 | 55 | 47 | 57 |
| 39 | 48 | 56 | 56 | 49 | 46 | 51 | 57 | 44 | 47 | 48 | 62 |
| 35 | 47 | 60 | 62 | 55 | 50 | 52 | 57 | 56 | 53 | 46 | 54 |
| 41 | 44 | 46 | 47 | 45 | 45 | 48 | 50 | 57 | 62 | 61 | 81 |
| 47 | 51 | 55 | 52 | 45 | 42 | 44 | 47 | 51 | 59 | 61 | 82 |
| 46 | 54 | 58 | 52 | 40 | 32 | 34 | 40 | 52 | 64 | 68 | 89 |
| 37 | 47 | 54 | 48 | 34 | 25 | 27 | 33 | 57 | 74 | 81 | 99 |

(c)

| 88 | 71 | 46 | 28 | 27 | 42 | 64 | 81 | 68 | 92 | 117 | 120 |
|----|----|----|----|----|----|----|----|----|----|-----|-----|
| 82 | 68 | 49 | 36 | 39 | 56 | 80 | 96 | 87 | 97 | 107 | 109 |
| 63 | 53 | 39 | 32 | 38 | 57 | 79 | 94 | 103 | 96 | 90 | 94 |
| 45 | 37 | 27 | 22 | 27 | 40 | 56 | 66 | 96 | 83 | 74 | 81 |
| 48 | 43 | 35 | 29 | 29 | 34 | 41 | 46 | 72 | 65 | 64 | 74 |
| 60 | 57 | 53 | 49 | 47 | 47 | 48 | 49 | 52 | 55 | 63 | 70 |
| 50 | 52 | 53 | 55 | 56 | 56 | 56 | 56 | 49 | 59 | 69 | 68 |
| 28 | 32 | 35 | 47 | 52 | 55 | 56 | 56 | 54 | 67 | 75 | 66 |
| 52 | 54 | 56 | 57 | 55 | 50 | 46 | 42 | 44 | 60 | 73 | 68 |
| 49 | 50 | 52 | 53 | 51 | 47 | 42 | 39 | 45 | 61 | 75 | 71 |
| 44 | 45 | 48 | 48 | 56 | 42 | 37 | 34 | 46 | 64 | 79 | 76 |
| 40 | 42 | 44 | 45 | 43 | 39 | 34 | 31 | 49 | 67 | 84 | 83 |

(d)

**FIGURE 8.3**   An example illustrating the property of PB features.

**TABLE 8.1**

**Position of a Pixel and Its Neighborhood at Different Compression Rates**

| Figure | Pixel Value | Sorted Sequence | Order |
|---|---|---|---|
| 8.3a | 34 | 28,31,32,33,*34*,36,40,46,64 | 5 |
| 8.3b | 35 | 20,27,31,*35*,39,42,48,49,66 | 4 |
| 8.3c | 32 | 24,27,32,*32*,38,42,43,45,59 | 4 |
| 8.3d | 39 | 22,27,32,36,37,*39*,49,53,68 | 6 |

(a)

| – | – | 1 | 1 | – | – |
|---|---|---|---|---|---|
| – | 1 | 2 | 2 | 1 | – |
| 1 | 2 | 1 | 1 | 3 | 1 |
| – | 2 | 1 | 2 | 3 | – |
| – | – | 1 | 2 | – | – |

(b)

| – | – | – | 1 | 1 | 1 | – | – | – |
|---|---|---|---|---|---|---|---|---|
| – | – | 1 | 2 | 2 | 2 | 1 | – | – |
| – | 1 | 2 | 1 | 3 | 1 | 2 | 1 | – |
| 1 | 2 | 2 | 2 | 3 | 3 | 3 | 2 | 1 |
| 1 | 2 | 3 | 2 | 4 | 3 | 3 | 2 | 1 |
| 1 | 2 | 2 | 3 | 3 | 3 | 3 | 2 | 1 |
| – | 1 | 2 | 2 | 3 | 2 | 2 | 1 | – |
| – | – | 1 | 2 | 2 | 2 | 1 | – | – |
| – | – | – | 1 | 1 | 1 | – | – | – |

(c)

| – | – | – | – | – | 1 | 1 | 1 | – | – | – | – | – |
|---|---|---|---|---|---|---|---|---|---|---|---|---|
| – | – | – | 2 | 2 | 2 | 2 | 2 | 2 | 2 | – | – | – |
| – | – | 3 | 3 | 3 | 3 | 3 | 3 | 3 | 3 | 3 | – | – |
| – | 2 | 3 | 4 | 4 | 4 | 4 | 4 | 4 | 4 | 3 | 2 | – |
| – | 2 | 3 | 4 | 5 | 5 | 5 | 5 | 5 | 4 | 3 | 2 | – |
| 1 | 2 | 3 | 4 | 5 | 6 | 6 | 6 | 5 | 4 | 3 | 2 | 1 |
| 1 | 2 | 3 | 4 | 5 | 6 | 7 | 6 | 5 | 4 | 3 | 2 | 1 |
| 1 | 2 | 3 | 4 | 5 | 6 | 6 | 6 | 5 | 4 | 3 | 2 | 1 |
| – | 2 | 3 | 4 | 5 | 5 | 5 | 5 | 5 | 4 | 3 | 2 | – |
| – | 2 | 3 | 4 | 4 | 4 | 4 | 4 | 4 | 4 | 3 | 2 | – |
| – | – | 3 | 3 | 3 | 3 | 3 | 3 | 3 | 3 | 3 | – | – |
| – | – | – | 2 | 2 | 2 | 2 | 2 | 2 | 2 | – | – | – |
| – | – | – | – | – | 1 | 1 | 1 | – | – | – | – | – |

**FIGURE 8.4** Circular GMSEs.

shown in Table 8.1. Using the PB features extraction algorithm and using T1 = 0.25 and T2 = 0.75, the pixel $p$ will receive 0 in the original and compressed images.

Rotation is one of the techniques to attack watermarking. In order to extract the PB features accurately and avoid rotational attack, we design the *grayscale morphological structuring element* (GMSE) to be circular. Figure 8.4 shows the circular GMSEs of different sizes. Apparently, the larger the GMSE is, the more circular it is. Note that the dash symbol (–) indicates "Don't care," which is equivalent to placing "$-\infty$" in the GMSE.

### 8.1.3 STRATEGIES FOR ADJUSTING VSTWs AND QFs

The size of the VSTW can determine whether the spatial or frequency domain is adopted. For example, if the size is $1 \times 1$, it is equivalent to the spatial domain approach. If the size is $2 \times 2$, $4 \times 4$, or $8 \times 8$, we split the original image using the sizes of the sub-blocks accordingly, and perform DCT on each sub-block. This

**TABLE 8.2**

**Relation between the QF and the Embedded Position of the Watermark**

| Int. | Binary | Embedded Watermark | Embedded Position | QF | After Embedding | New Int. |
|------|--------|--------------------|-------------------|-----|-----------------|----------|
| 41 | 0010100*1* | 0 | 1 (LSB) | 1 | 00101000 | 40 |
| 43 | 0010*1*011 | 0 | 4 | 8 | 00100011 | 35 |
| 66 | 010000*1*0 | 0 | 2 | 2 | 01000000 | 64 |
| 110 | 0110*1*110 | 0 | 4 | 8 | 01100110 | 102 |
| 210 | 110*1*0010 | 0 | 5 | 16 | 11000010 | 194 |

belongs to the frequency domain approach. Note that the size of the VSTW could be arbitrary, not necessarily a square to the power of 2.

In principle, fragile watermarks are embedded into the LSBs, but robust watermarks are embedded into more significant bits. In order to integrate both watermarks, the QFs are developed based on the spread spectrum. We know that shifting the binary bits of $x$ to the left (denoted as $shl$) or the right (denoted as $shr$) by $y$ bits is equivalent to multiplying or dividing $x$ by $2^y$. The position for embedding the watermark can be determined based on the QF, as shown in Table 8.2. For example, if the watermark is embedded into the fourth bit (i.e., QF = 8) of a pixel, we will divide the pixel value by 8, and then replace the resulting LSB with the watermark. Finally, we multiply the result by 8 to obtain the watermarked value.

Cox et al. [1] proposed spread-spectrum watermarking and indicated that the watermarks should not be embedded into insignificant regions of the image or its spectrum since many signal and geometric processes can affect these components. For robustness, they suggested that the watermarks should be embedded into the regions with large magnitudes in the coefficients of a transformed image. For example, there are a few large magnitudes in the coefficients shown in Figure 8.5. Figure 8.5a shows the original image of size 16 × 16, in which all pixels are 210. Figure 8.5b is the image transformed by the DCT using the 8 × 8 VSTW. There are four significant regions with a large magnitude. Figure 8.5c shows the results after embedding the watermarks into the sixth bit of four different locations, as marked in a bold box. Note that the watermarks are not embedded into the four significant regions. After performing the IDCT on Figure 8.5c, we obtain the watermarked image shown in Figure 8.5d. It is obvious that the original image will suffer from huge degradation because the watermarks are embedded into nonsignificant regions. Therefore, the size of the embedded watermarks is quite limited.

Figure 8.6 shows an example of our AP watermarking for enlarging the capacity of the watermarks. In order to achieve this, a small-sized VSTW is selected. The watermark uses the 64 bits that are all 1s. Considering robustness, the watermark is embedded into the most significant bit (e.g., the fifth bit). Given the size of the VSTW is 8 × 8, we obtain Figure 8.6a by embedding the watermark into Figure 8.5b. After performing the IDCT on Figure 8.6a, the watermarked image is obtained, as shown

(a)

(b)

(c)

(d)

**FIGURE 8.5** An example of embedding watermarks into different spectral regions.

(a)

(b)

(c)

(d)

FIGURE 8.6    An example of enlarging the capacity of watermark.

**TABLE 8.3**

**General Rules for Determining the VSTW and QF**

| Purpose | VSTW | QF |
|---|---|---|
| Robust and spatial | $1 \times 1$ | $> 1$ |
| Robust and frequency | $> 1 \times 1$ | $> 1$ |
| Fragile and spatial | $1 \times 1$ | $\leq 1$ |
| Fragile and frequency | $> 1 \times 1$ | $\leq 1$ |

in Figure 8.6b, which is largely distorted. It is obvious that the performance is not ideal since too many data are embedded. However, if the $2 \times 2$ VSTW is selected we can obtain Figure 8.6c, which is the result of embedding the watermark into the frequency domain of the original image. Similarly, after performing the IDCT on Figure 8.6c, we can obtain the watermarked image shown in Figure 8.6d. Therefore, using the VSTW, we can determine the size of each subimage to enlarge the capacity of the watermarks.

By adjusting the VSTW and QF, our technique can be equivalent to existing watermarking techniques. For example, by setting the VSTW to be $N \times N$ and each element in the QF to be a large number, our AP watermarking technique achieves the same results as the spread-spectrum technique developed by Cox et al. [1]. To simulate Wong's block-based fragile watermarking [2], we set the VSTW to be $1 \times 1$ and each element in the QF to be 1. The general rules for determining the VSTW and QF are shown in Table 8.3.

### 8.1.4 EXPERIMENTAL RESULTS

Figures 8.7a,b show an original image and its PB features, respectively. Figures 8.7c,d show the image after JPEG compression with a 20% quality level and its PB features, respectively. By comparing Figures 8.7b,d, the PB features almost remain the same even after low-quality JPEG compression. Table 8.4 provides the quantitative measures. Note that the size of the structuring element is $3 \times 3$ and $(T_1, T_2)$ indicates the threshold values adopted in step 4 of our algorithm. Table 8.5 provides an analysis based on different sizes of the structuring element.

Figures 8.8a,b show an image and its data of size $16 \times 16$. The size of the VSTW adopted is $4 \times 4$ in this example. Figure 8.8c shows the QF corresponding to each subimage. Figure 8.8f is the original watermark. Note that the watermark is embedded into the DC components of the coefficients in the transformed subimages. That is, one bit of the watermark is inserted in each subimage. Figures 8.8d,e show the watermarked image and its data. JPEG compression is adopted as the attacking method. The pairs in Figure 8.8—(g,h), (j,k), (m,n), and (p,q)—are the results of JPEG compression at four quality levels: 80%, 60%, 40%, and 20%, respectively. Figures 8.8i,l,o,r are the corresponding extracted watermarks.

Figure 8.9 shows an example of our AP watermarking algorithm for extracting the embedded watermarks after different JPEG compression qualities. Figures 8.9 a1–c1 are the original watermarks of sizes $128 \times 128$, $64 \times 64$, and $32 \times 32$,

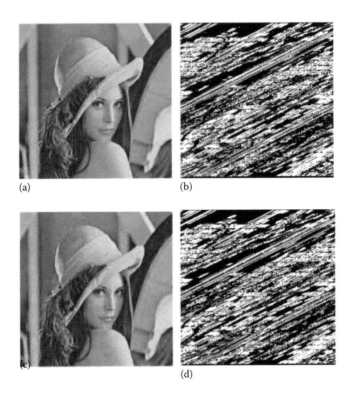

(a)    (b)

(c)    (d)

**FIGURE 8.7**  An example of the PB features of an image.

respectively. The sizes of VSTWs used for embedding those watermarks are $2 \times 2$, $4 \times 4$, and $8 \times 8$, and their corresponding QFs are shown in Figures 8.10a–c. After embedding the watermarks into the host images, we use four JPEG compression qualities—80%, 60%, 40%, and 20%—as the attackers. Figures 8.9 a2–c2 are the extracted watermarks after 80% JPEG compression. Similarly, Figures 8.9a3–c3, a4–c4, and a5–c5 are obtained after 60%, 40%, and 20% compression, respectively. Figures 8.8 and 8.9 demonstrate that our AP watermarking technique can resist JPEG compression attacks.

### 8.1.5 COLLECTING APPROACH TO GENERATING VSTWS

In order to extend our AP technique to general images whose size is not quadrate, we develop a collecting approach to generating VSTWs. Unlike the traditional approach of separating an original image into lots of non-overlapping blocks, the collecting approach systematically picks up the proper pixels to form the VSTW. Figure 8.10 shows the QFs for their corresponding VSTWs. Figure 8.11 illustrates the collecting approach, where (a) is the original image, and (b)–(d) show the collection of 4, 16, and 64 pixels to generate VSTWs of size $2 \times 2$, $4 \times 4$, and $8 \times 8$, respectively. Due to the space correlation, most coefficients of the transformed blocks obtained by the traditional approach are 0s, except the low-frequency parts. Our collecting approach

**TABLE 8.4**

**Quantitative Measures for the PB Features of an Image after Different Levels of JPEG Compression**

| JPEG Quality | PSNR | $T_1 = 0.1\, T_2 = 0.9$ | $T_1 = 0.15\, T_2 = 0.85$ | $T_1 = 0.2\, T_2 = 0.8$ | $T_1 = 0.25\, T_2 = 0.75$ | $T_1 = 0.3\, T_2 = 0.7$ | $T_1 = 0.35\, T_2 = 0.65$ |
|---|---|---|---|---|---|---|---|
| 90% | 48.98 | 98.67% | 98.31% | 94.14% | 98.05% | 96.78% | 95.73% |
| 80% | 44.92 | 96.44% | 95.98% | 95.78% | 95.67% | 95.04% | 93.21% |
| 70% | 41.99 | 95.88% | 95.09% | 94.84% | 94.65% | 93.77% | 91.84% |
| 60% | 39.18 | 95.45% | 94.91% | 94.69% | 94.20% | 93.50% | 91.31% |
| 50% | 38.68 | 94.64% | 94.04% | 93.67% | 93.37% | 92.33% | 90.20% |
| 40% | 35.81 | 94.15% | 93.48% | 93.08% | 92.49% | 92.12% | 90.06% |
| 30% | 32.33 | 92.22% | 91.24% | 90.73% | 90.42% | 89.62% | 86.44% |
| 20% | 30.57 | 91.45% | 90.38% | 89.93% | 89.38% | 88.47% | 86.78% |
| 10% | 24.51 | 74.11% | 74.08% | 73.96% | 74.15% | 75.09% | 78.73% |

**TABLE 8.5**
**PB Features Based on Sizes of the Structuring Element**

| | SE = 5 × 5 | | SE = 7 × 7 | |
|---|---|---|---|---|
| JPEG Quality | $T_1 = 0.1\,T_2 = 0.9$ | $T_1 = 0.25\,T_2 = 0.75$ | $T_1 = 0.1\,T_2 = 0.9$ | $T_1 = 0.25\,T_2 = 0.75$ |
| 90% | 98.75% | 98.27% | 99.10% | 98.68% |
| 80% | 98.15% | 98.27% | 98.56% | 98.98% |
| 70% | 98.38% | 96.05% | 98.71% | 96.67% |
| 60% | 98.16% | 95.87% | 96.65% | 96.35% |
| 50% | 96.30% | 94.60% | 96.52% | 95.55% |
| 40% | 95.92% | 94.43% | 96.38% | 95.06% |
| 30% | 94.32% | 91.73% | 95.04% | 93.18% |
| 20% | 93.90% | 91.76% | 94.37% | 92.83% |
| 10% | 76.89% | 76.78% | 79.30% | 79.18% |

will break the space correlation. Therefore, it not only extends our AP technique to general-sized images but also enlarges the capacity of watermarks when the frequency domain approach is used.

## 8.2    ADJUSTED-PURPOSE WATERMARKING USING PARTICLE SWARM OPTIMIZATION

Image watermarking has been adopted as an efficient tool for identifying the source, creator, distributor, or authorized consumer of an image, as well as for detecting whether an image has been illegally altered. There are several critical factors to be considered, including imperceptibility, capacity, and robustness. An ideal watermarking system should be able to embed a large amount of information perfectly securely without visible degradation to the host image [10]. In other words, the perceptual similarity between the host image and the stego-image (i.e., the image after embedding the watermark) must be sufficiently high while a large amount of information is embedded.

Basically, there are two image domains for embedding a watermark: one is the spatial domain and the other is the frequency domain. Watermarking in the spatial domain is to insert data into a host image by changing the gray levels of certain pixels in the host image [1–14]. It has the advantages of low computational cost and high capacity; however, the inserted information may be less robust and can be easily detected by computer analysis. On the other hand, watermarks can be embedded into the coefficients of a transformed image in the frequency domain [13]. The transformation includes the *discrete cosine transform* (DCT), *discrete Fourier transform* (DFT), or *discrete wavelet transform* (DWT). In this way, the embedded data are distributed around the entire image, which makes the watermark difficult to detect. However, the embedding capacity of the frequency domain is often lower than that of the spatial domain since the image quality will be significantly degraded if the amount of embedding data is enlarged. Therefore, a practical watermarking scheme

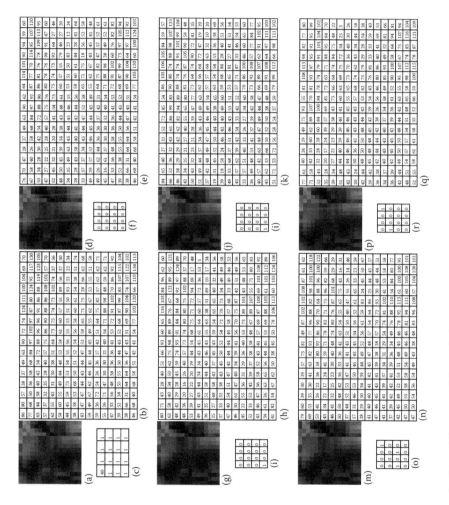

**FIGURE 8.8** An example of our AP watermarking technique.

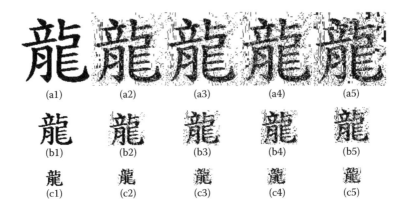

|     |     |     |     |     |
|-----|-----|-----|-----|-----|
| (a1) | (a2) | (a3) | (a4) | (a5) |
| (b1) | (b2) | (b3) | (b4) | (b5) |
| (c1) | (c2) | (c3) | (c4) | (c5) |

**FIGURE 8.9**    An example of extracting embedded watermarks after JPEG compression.

| 16 | 1 | 1 | 1 | 1 | 1 | 1 | 1 |
|----|---|---|---|---|---|---|---|
| 1  | 1 | 1 | 1 | 1 | 1 | 1 | 1 |
| 1  | 1 | 1 | 1 | 1 | 1 | 1 | 1 |
| 1  | 1 | 1 | 1 | 1 | 1 | 1 | 1 |
| 1  | 1 | 1 | 1 | 1 | 1 | 1 | 1 |
| 1  | 1 | 1 | 1 | 1 | 1 | 1 |   |
| 1  | 1 | 1 | 1 | 1 | 1 | 1 | 1 |
| 1  | 1 | 1 | 1 | 1 | 1 | 1 | 1 |

(a)

| 16 | 1 | 1 | 1 |
|----|---|---|---|
| 1  | 1 | 1 | 1 |
| 1  | 1 | 1 | 1 |
| 1  | 1 | 1 | 1 |

(b)

| 16 | 1 |
|----|---|
| 1  | 1 |

(c)

**FIGURE 8.10**    QFs and their corresponding VSTWs.

should integrate the characteristics of high embedding capacity and high stego-image quality. As for watermarking robustness, we prefer to embed the watermark into the frequency domain.

Taking advantage of embedding a watermark in the difference expansion [15–17] of the spatial domain, we embed secret data into the magnitude of the cosine harmonic waves of a host image in the frequency domain using the DCT. Then, the DWT and *singular value decomposition* (SVD) are applied to enhance the robustness [18,19], since SVD preserves both the one-way and nonsymmetrical properties of an image. Therefore, to achieve the goal of increasing the embedding capacity, we have developed a DWT–SVD–DCT-based embedding algorithm. The reversible ability [15,17,20–22] is considered in the proposed algorithm to recover the original host image.

To define *regions of interest* (ROIs) for quality preservation automatically, we have developed a method to select the areas containing the most important information in an image by using saliency detection to preserve proto-objects as ROIs. Salient areas are regions that stand out from their neighborhoods [11,12,14]. A saliency detection model based on the spectral residuals of the magnitude spectrum after the Fourier transform was used by Hou and Zhang [12]. Guo et al. [11] pointed out that salient areas should come from the phase spectrum instead of the magnitude spectrum.

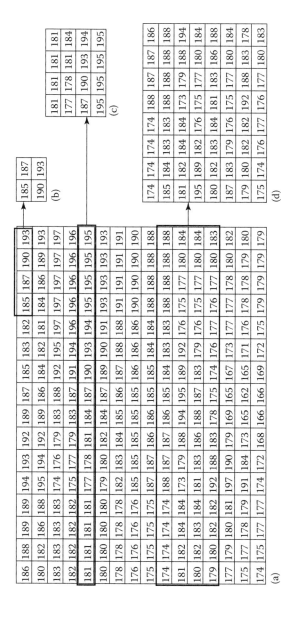

**FIGURE 8.11** An example of the collecting approach to generating the VSTW.

Given a host image and a watermark image, we intend to develop an optimal watermarking solution considering imperceptibility, capacity, and robustness by using *particle swarm optimization* (PSO) [23]. PSO is a population-based stochastic optimization technique inspired by the social behavior of bird flocking or fish schooling. It can optimize image-watermarking schemes [19] when the embedding objectives are contradictory.

Our previous work [24] focused on single-ROI medical image watermarking and the correction of rounding errors, which are deviations in the conversion of real numbers into integers. In this chapter, we extend our research to extracting multiple ROIs as well as automatic selection and propose a novel scheme of RONI ranking to achieve high-capacity watermark embedding in the frequency domain and high-quality stego-images. Different amounts of bits in watermarks can be adjusted for different purposes based on PSO determination.

The section is organized as follows: Section 8.2.1 presents the proposed technique, and experimental results and analysis are described in Section 8.2.2.

### 8.2.1 Proposed Technique

In order to increase the embedded watermark capacity without distorting the important information in an image, we automatically select multiple ROIs in an image to be lossless and the residual *regions of noninterest* (RONIs) to be lossy. The RONIs constitute a concave rectangle. The proposed technique is divided into four steps. First, the ROIs and RONIs are separated through proto-object detection based on a saliency map. Second, the concave RONI rectangle is decomposed into a minimum number of non-overlapping rectangles through partitioning, which are then respectively transformed into the frequency domain. Third, we perform our embedding algorithm on the preprocessed watermark information bitstream along with the amount to be embedded, determined by the PSO of each RONI rectangle, to generate an optimal embedded rectangle and a corresponding key. Fourth, the stego-image is obtained by combining those embedded rectangles and the ROIs. Figure 8.12 shows the overall embedding process.

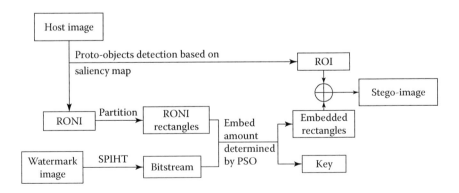

**FIGURE 8.12**   The overall embedding process.

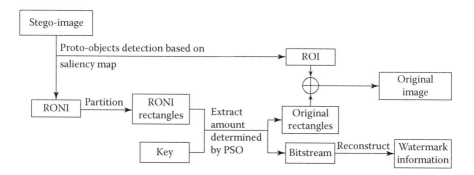

**FIGURE 8.13**   The overall decoding process.

During the decoding process, we obtain the ROIs and RONIs from the stego-image, collate RONI rectangles, and perform our decoding algorithm with the amount to be extracted, which is determined by the PSO along with a key to extract the original rectangle and a bitstream. We obtain the original image by combining the generated rectangles and ROIs, and retrieve the original watermark by reconstructing the bitstream. Figure 8.13 shows the overall decoding process.

### 8.2.1.1   ROI Automatic Extraction

For the purpose of automatic ROI selection, the areas containing the most important information for the human visual system must be defined. Preserving those areas as ROIs will avoid watermark embedding and ensure no modification. We select the proto-objects generated from the saliency map as ROIs. The saliency map is obtained by the Fourier transform. Since images follow the power law [25,26], most of the energy comes from the low-frequency areas. In other words, the magnitude spectrums of different images after the Fourier transform share a similar shape. Since there are only magnitude spectrums and phase spectrums after the Fourier transform, the magnitude spectrums are similar and the saliency or major differences in images should come from the phase spectrum. Medical images, especially, can be viewed as piecewise smooth.

The saliency map of an image is obtained by removing the magnitude spectrum and reconstructing the image using only the phase spectrum after the Fourier transform. The process can be mathematically stated as

$$\text{Map} = \mid \text{IFT}\big(\exp\{j * \text{phase}(\text{FT}(I))\}\big)\mid, \tag{8.1}$$

where $I$ is an input image, FT denotes the Fourier transform, "phase" means the phase spectrum, $j$ is $\sqrt{-1}$, and IFT denotes the inverse Fourier transform. For simplicity, the saliency map can be obtained by using a blurring filter—for example, a Gaussian filter or average filter.

The proto-object mask is generated by a threshold that is set empirically as a constant multiplied by the arithmetic mean value of the saliency map. For natural images, the constant is between 1 and 4. Figure 8.14 shows the result of the proto-object

**FIGURE 8.14** Automatic extraction of ROIs.

extraction for a natural image. Note that for conducting watermark embedding in a block, we choose the smallest bounding rectangular shape for each ROI.

### 8.2.1.2 Watermark Preprocessing

The purpose of preprocessing watermark information is to generate a bitstream for embedding. For watermark images we use bit planes, and for textual data we convert them into images using encoding techniques such as barcoding or QR coding.

Given a watermark image, each pixel is represented by eight bits, where usually the first five bits are considered the most significant and the last three bits are less important. The bitstream of the watermark is generated by combining all the bits of entire pixels. The reconstruction process is to convert the bitstream back into numbers, with eight bits being combined into an integer.

### 8.2.1.3 RONI Partitioning

Partitioning RONIs involves decomposing an image into a number of RONI non-overlapping rectangular blocks, which are transformed from the spatial to the frequency domain. Given a rectilinear concave polygon $M$, the minimum number of rectangles $P$ that $M$ can be partitioned into is computed by

$$P = \frac{n}{2} + h - 1 - c. \tag{8.2}$$

where:
- $n$    denotes the number of vertices of $M$
- $h$    is the number of holes of $M$
- $c$    denotes the maximum cardinality of a set $S$ of concave lines, no two of which are intersected [27]

We have developed a novel algorithm for partitioning a RONI concave rectangle into a minimum number of rectangles without the coordinates of all vertices. This algorithm first draws horizontal and vertical concave lines for each ROI, respectively, and then selects the result of the minimum rectangles as the partition.

Let $[(x_1,y_1), (x_2,y_2), (x_3,y_3), (x_4,y_4)]$ denote the coordinates of the upper-left, upper-right, lower-left, and lower-right corners of each ROI, respectively, with the origin located on the upper-left corner of the image. The RONI concave rectangle partitioning algorithm is presented as follows.

*Algorithm 1: RONI concave rectangle partitioning*
Input: Image, the ROIs of which are the coordinates $[(x_1,y_1), (x_2,y_2), (x_3,y_3), (x_4,y_4)]$
Output: Partition of minimum rectangles

1. For each ROI, do
   a. Keep decreasing $x_1,x_3$ by one pixel until hitting the margin or another ROI, and record $x_{1\_end}, x_{3\_end}$.
   b. Keep increasing $x_2,x_4$ by one pixel until hitting the margin or another ROI, and record $x_{2\_end}, x_{4\_end}$.
   c. Keep decreasing $y_1,y_2$ by one pixel until hitting the margin or another ROI, and record $y_{1\_end}, y_{2\_end}$.
   d. Keep increasing $y_3,y_4$ by one pixel until hitting the margin or another ROI, and record $y_{3\_end}, y_{4\_end}$.
   End for each:
2. Draw lines from point $(x_{1\_end}, y_1)$ to point $(x_{2\_end}, y_2)$ and from point $(x_{3\_end}, y_3)$ to point $(x_{4\_end}, y_4)$ to generate a horizontal partitioned result.
3. Draw lines from point $(x_1, y_{1\_end})$ to point $(x_3, y_{3\_end})$ and from point $(x_2, y_{2\_end})$ to point $(x_4, y_{4\_end})$ to generate a vertical partitioned result.
4. Return the result of the generated partition rectangles as the final partitioned result.

Figure 8.15 shows an example of partitioning a RONI concave rectangle into a minimum number of rectangles. Figure 8.15a shows an original natural image with three ROIs extracted. Figure 8.15b shows 10 white rectangles, which are the corresponding RONI partitioned rectangles. According to Equation 8.2, we have $n = 16$, $h = 3$, and $c = 0$, and therefore $P = 10$.

## 8.2.1.4 Optimal Watermarking Scheme Using PSO

In this section, we develop a new optimal watermarking scheme for embedding high-capacity data by PSO. Since PSO can iteratively optimize a problem by improving

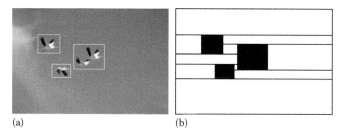

(a)                                (b)

**FIGURE 8.15** An example of RONI partitioning: (a) a three-ROI image, (b) RONI rectangles.

existing solutions, each solution is a particle moving in the search space at each round. We divide this algorithm into four steps as follows:

1. Randomly initialize the position and velocity of particles.
2. Compute the new position $P'$ and the new velocity $V'$ by

$$P' = P + mV, \tag{8.3}$$

$$V' = V + (c_1 \times rand \times (G - P)) + (c_2 \times rand \times (PB - P)), \tag{8.4}$$

where:
   $P$   is current position of a particle
   $PB$   is the best position of current particle at current iteration
   $rand$ is a random number
   $G$   is the best particle at current iteration
   $V$   is current velocity of a particle
   $m, c_1,$ and $c_2$ are the updating pace factors

We set $m = 0.1$, $c_1 = 1$, and $c_0 = 1$ according to Kennedy's suggestion [23].

3. Keep refreshing the best particle at current iteration $G$ and the best position of current particle $PB$ according to the fitness function.
4. Terminate the program until the predefined conditions have been satisfied.

The fitness function is designed to achieve higher quality and higher capacity after inserting a watermark image into a host image. We consider the universal image quality index [28] for quality preservation and the number of bits for capacity counting. The fitness function is computed as

$$\text{fitness} = w_1 Q + w_2 C. \tag{8.5}$$

Considering a host image $I_1$ and a stego-image $I_2$ of size $M \times N$, the universal image quality matrix $Q$ can be computed by

$$Q = \frac{\sigma_{I_1 I_2}}{\sigma_{I_1} \sigma_{I_2}} \times \frac{2\overline{I_1} \overline{I_2}}{\overline{I_1}^2 + \overline{I_2}^2} \times \frac{2\sigma_{I_1}\sigma_{I_2}}{\sigma_{I_1}^2 + \sigma_{I_2}^2}, \tag{8.6}$$

where:

$$\overline{I_1} = \frac{1}{MN} \sum I_1 \tag{8.7}$$

$$\overline{I_2} = \frac{1}{MN} \sum I_2 \tag{8.8}$$

$$\sigma_{I_1 I_2} = \sum \sum (I_1 - \overline{I_1})(I_2 - \overline{I_2}) \tag{8.9}$$

$$\sigma_{I_1}^{2} = \sum\sum (I_1 - \overline{I_1})^2 \tag{8.10}$$

$$\sigma_{I_2}^{2} = \sum\sum (I_2 - \overline{I_2})^2 \tag{8.11}$$

The first term of $Q$ is the correlation coefficient, which measures the correlation loss between $I_1$ and $I_2$. The range is $(-1,1)$, with the best value 1 indicating that both standard deviations are the same. The second term measures the luminance loss, and the range is $(0,1)$, with the best value 1 indicating both mean luminances are the same. The third term measures the contrast loss, and the range is $(0,1)$, with the best value 1 indicating no loss of contrast. Overall, $Q$ is in the range of $(-1,1)$, with the best value 1 indicating images $I_1$ and $I_2$ are identical to each other. $C$ is the number of bits (i.e., capacity) to be embedded in $I_1$.

There are two weighting factors, $w_1$ and $w_2$, of the fitness function in Equation 8.5. They respectively determine the importance of fidelity and capacity when a watermark is embedded to make our scheme adjustable in purpose. For instance, if quality preservation is the primary purpose, $w_1$ could be set larger and $w_2$ could be set smaller. Note that $w_1, w_2 \in [0,1]$, and $w_1 + w_2 = 1$.

### 8.2.1.5 Algorithm for Embedding and Extracting

As discussed in Section 8.2.1.1, most image differences come from the phase spectrum after the Fourier transform, while the magnitude spectrum of different images shares a similar shape. We exploit the phase spectrum to obtain the crucial ROIs and utilize the magnitude spectrum for embedding secret data. The overall magnitude spectrum shape is maintained, while the hidden data only weaken the magnitude values. Hence, the magnitude of the harmonic waves of DCT coefficients is adopted.

For each partitioned rectangle, we transform it using DWT, decompose the horizontal (HL) band using SVD, compute the principal components, and transform the principal components using DCT. The *middle frequency area* (MFA) after DCT has been chosen for embedding because the MFA avoids the most important parts for human vision (i.e., low-frequency components) while significantly protecting the watermark from removal due to compression or noise (i.e., high-frequency components).

After MFA selection, we perform the embedding algorithm for the bitstream. The magnitude in the MFA will be slightly weakened after embedding. The number of embedding loops $t$, generated by PSO, determines the weakening level. The detailed embedding algorithm is presented as Algorithm 2. Note that if rounding errors occur in the process of the DCT, we can correct them by applying our previous work in [24].

*Algorithm 2: The embedding algorithm*
Input: Host rectangle $H$, bitstream $S$, and the number of embedding loops $t$
Output: Embedded rectangle and a key $K$

1. Perform the DWT on $H$ and choose the $HL$ band.
2. Perform SVD on the $HL$ band and compute the principal component *PHL* by

$$PHL = UHL * SHL, \qquad (8.12)$$

where *UHL* and *SHL* are obtained by

$$[UHL, SHL, VHL] = \text{SVD}(HL). \qquad (8.13)$$

3. Perform DCT on the *PHL* and select the MFA.
4. Obtain the mean value *M* of the MFA.
5. For each coefficient *V* in the MFA, compute middle point *A* by calculating the arithmetic mean of *V* and *M*. Store *A* as the key *K*. Note that for further security concerns, we could apply an encrypting algorithm on *K*.
6. Compute the embedding length *L* by

$$L = |\text{round}(\log |A - M|)|. \qquad (8.14)$$

Note that if $L = 0$ or $|A - M| <$ the logarithm base, one bit will be embedded.
7. Obtain the embedding integer value *E* by taking *L* bits of *S*.
8. If *V* is positive, update $V = V + E$;
   else update $V = V - E$.
9. Repeat steps 5 through 8 *t* times.

Figure 8.16 shows the process of embedding the watermark. Finally, all the embedded RONI rectangles are combined with the ROIs to form the stego-image.

For watermark extraction, we select a RONI watermarked rectangle, perform the reverse extracting algorithm, and combine all the decoded rectangles to obtain the original image. Note that the ROI mask was included as part of the key in embedding. Figure 8.17 shows the detailed decoding process. The detailed decoding algorithm is presented as Algorithm 3. Note that we can retrieve the watermark with the reconstruction algorithm described in Section 8.2.1.2.

*Algorithm 3: The extracting algorithm*
Input: Watermarked rectangle *H′*, key *K*, and the number of embedding loops *t*
Output: Original rectangle and the watermark bitstream *B*

1. Perform the DWT on *H′* and choose the HL band, called *HL′*.
2. Perform SVD on *HL′* and compute the principal components, called *PHL′*.
3. Perform DCT on *PHL′* and select the middle frequency area, called *MFA′*.
4. Obtain the mean value *M′* of *MFA′*.
5. For each coefficient *V′* in *MFA′*, we compute the embedded value *E′* by

$$E' = |V' - K|. \qquad (8.15)$$

6. Compute the embedded length L′ by

$$L' = |\text{round}(\log |K - M'|)|. \qquad (8.16)$$

Note that if $L' = 0$ or $|K - M'| <$ the logarithm base, one bit will be extracted.

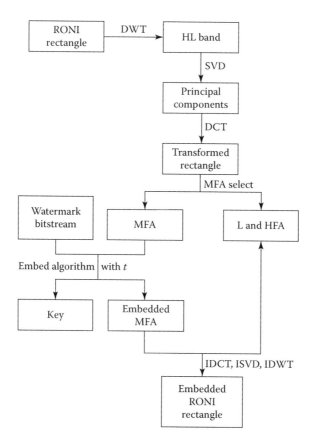

**FIGURE 8.16**    The detailed embedding algorithm.

7. Obtain the embedded bitstream $B'$ by converting $E'$ with length $L'$.
8. For each coefficient $V'$ in the $MFA'$, we update $V'$ by

$$V' = 2K - M'. \qquad (8.17)$$

9. Obtain the original $MFA$ by repeating steps 5 to 8 $t$ times and obtain the whole bitstream $B$ by stacking B'.

## 8.2.2  EXPERIMENTAL RESULTS

Figure 8.18 shows an example of embedding a sample watermark image into a natural image. Figure 8.18a shows a natural image of size $500 \times 300$ with four automatically generated ROI. Figure 8.18b is the corresponding RONI partition, with 13 rectangles in this case. Figure 8.18c shows the watermark logo. Figure 8.18d is the stego-image with a *peak signal-to-noise ratio* (PSNR) of 57.33. Figure 8.18e is the difference between the original and stego-images. Note that we amplify the differences 100

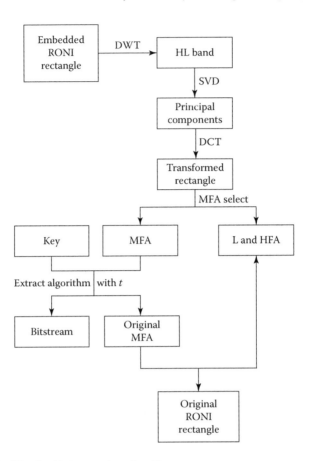

**FIGURE 8.17**   The detailed extracting algorithm.

times for highlighting. We can see that the four ROIs are lossless. In this case we set the quality and the capacity to be equally important; that is, $w_1 = w_2 = 0.5$.

The embedding result of this example is listed in Table 8.6. In this case, only the first seven RONI rectangles are used for embedding, as shown in Figure 8.19. The remaining RONI rectangles are available, but not used for watermark embedding due to the size of this watermark image. Note that $t$ is the number of embedding loops. For each of the seven RONI rectangles, $t$ is obtained by the proposed algorithm to be 6, 5, 6, 6, 6, 5, and 4, respectively, in a column-wise order. The embedding bitstream length is the number of bits for embedding. The *used area* means the number of pixels used for embedding. The PSNR between a host image $f$ and a stego-image $g$ is computed by

$$\mathrm{PSNR}(f,\ g) = 10 \times \log_{10} \frac{(2^k - 1)^2}{MSE}, \tag{8.18}$$

where $k$ is the bit depth and *MSE* denotes mean square error, which is obtained by

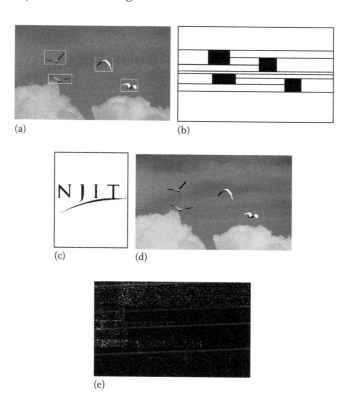

**FIGURE 8.18** An example of embedding a sample watermark image: (a) a natural image with ROI detection, (b) RONI rectangle partition, (c) the watermark logo image, (d) the stego-image, (e) the difference between original and stego-images (amplified 100 times).

**TABLE 8.6**
**Bitstream Length, Embed Area, PSNR, and Q**

| Bitstream Length (bits) | Used Area | PSNR | Q |
|---|---|---|---|
| 52,000 | 9,968 | 57.33 | 0.99 |

$$MSE = \frac{1}{mn} \sum_{1}^{m} \sum_{1}^{n} \| f(i,j) - g(i,j) \|^2. \tag{8.19}$$

Table 8.6 implies that the bits per pixel (BPP) in Figure 8.19 is 5.22, which is computed as the total bits divided by the used area. We can increase $t$ to achieve a higher BPP; however, the PSNR would decrease correspondingly. Note that our experiment is conducted using grayscale natural images. For color images, we can apply the proposed scheme on the luminance channel.

Since we have multiple RONI partitioned rectangles and a different $t$ for each rectangle, it is difficult to explore the relationship as a whole. For illustration purposes,

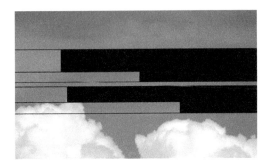

**FIGURE 8.19**   The RONI rectangles used for embedding.

**TABLE 8.7**
**PSNR, BPP, and Capacity (Number of Embedding Loops *t* from 1 to 6)**

| *t* | 1 | 2 | 3 | 4 | 5 | 6 |
|---|---|---|---|---|---|---|
| PSNR | 52.00 | 45.17 | 38.83 | 32.46 | 36.27 | 20.15 |
| Q | 0.99 | 0.90 | 0.66 | 0.56 | 0.53 | 0.52 |
| BPP | 1.77 | 3.73 | 6.18 | 9.39 | 13.55 | 18.70 |

we take the first rectangle out and compute its PSNR, Q, and BPP. The result of the embedded area is 5274 pixels, and *t* ranges from 1 to 6, as shown in Table 8.7.

Figure 8.20a plots the relationship curve between *t* and PSNR, Figure 8.20b plots the relationship curve between *t* and BPP, and Figure 8.20c plots the relationship curve between *t* and Q. We can observe that as *t* increases, BPP increases; however, PSNR and Q decrease. The relationship is similar for all other RONI rectangles. Note that, more specifically, we can adjust a different ratio of $w_1$ and $w_2$ to emphasize image quality or watermark capacity based on the user's applications.

Recall that $w_1$ and $w_2$ are the weighting factors of quality and capacity respectively, and $w_1, w_2 \in [0,1]$ and $w_1 + w_2 = 1$. Let *t* be limited from 1 to 6. First, to emphasize embedding capacity, $w_1$ is set to be 0.1 and $w_2$ is set to be 0.9. As a result, we obtain $t = 6$, PSNR = 20.15, and BPP = 18.70. High capacity is achieved as the embedding purpose. Second, to emphasize embedding quality, $w_1$ is set to be 0.9 and $w_2$ is set to be 0.1. As a result, we obtain $t = 2$, PSNR = 45.71, and BPP = 3.73. High quality is also achieved as the embedding purpose. Note that in this example only one rectangle of the host image is used due to the limited size of the watermark image. However, in some applications using the entire rectangles of the host image, quality or fidelity will be further enhanced due to the preservation of ROIs.

From experimental results, it is observed that our proposed algorithm has significant improvements over currently available techniques, as well as our previous work in [24] of BPP = 1 and an average PSNR around 38 with a single ROI. Therefore, the proposed scheme allows the automatic detection of multiple ROIs for quality preservation and uses DWT–SVD imperceptibility, DCT magnitude expansion for large capacities, and PSO for optimal embedding. For the number of embedding loops *t*

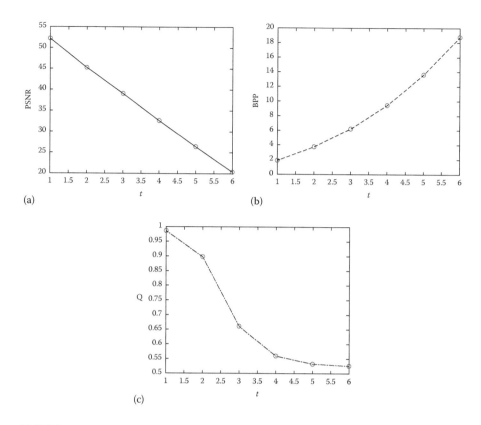

(a)

(b)

(c)

**FIGURE 8.20** Plots for *t* from 1 to 6.

from 1 to 6, PSNR can be improved by up to 37% (with the best case of $t = 1$) and BPP is improved up to 1770% (with the best case of $t = 6$). It is a trade-off between capacity and quality, and PSO intelligently and automatically solves the problem.

The computational complexity of the proposed algorithm is analyzed as follows. Most evolutionary algorithms have, at each iteration, a complexity of $O(n*p + Cof*p)$, where $n$ is the dimension of the problem, $p$ is the population size, and $Cof$ is the cost of the objective function. Furthermore, assume that an evolutionary algorithm performs $FEs/p$ iterations, where $FEs$ is the maximum amount of function evaluations allowed. Thus, the complexity cost becomes $O(n*FEs + Cof*FEs)$. The second term tends to dominate the time complexity, and this complexity is determined by the cost of evaluating the objective function and the amount of evaluations performed. Therefore, the algorithm complexity is measured by the amount of evaluations it performs.

In summary, the proposed technique has three advantages. First, it keeps the automatically detected ROIs lossless. The proposed technique carefully preserves the same image quality for those regions crucial to human vision. Second, the proposed technique enhances the capacity over the traditional frequency domain methods. Third, given a host image and a watermark image, the proposed algorithm provides an optimal watermarking solution, and for specific purposes the emphasis of capacity or quality is adjustable.

## REFERENCES

1. Cox, I. et al., Secure spread spectrum watermarking for multimedia, *IEEE Trans. Image Processing*, 6, 1673, 1997.
2. Wong, P. W., A public key watermark for image verification and authentication, in *Proc. IEEE Int. Conf. Image Processing*, Chicago, IL, 1998, 425.
3. Rivest, R. L., Shamir, A., and Adleman, L., A method for obtaining digital signatures and public-key cryptosystems, *Comm. ACM*, 21, 120, 1978.
4. Rivest, R. L., The MD5 message digest algorithm, RFC 1321, accessed on January 6, 2017, http://www.faqs.org/rfcs/rfc1321.html, April 1992.
5. Celik, M. U. et al., Hierarchical watermarking for secure image authentication with localization, *IEEE Trans. Image Processing*, 11, 585, 2002.
6. Chen, L.-H. and Lin, J.-J., Mean quantization based image watermarking, *Image Vision Computing*, 21, 717, 2003.
7. Serra, J., *Image Analysis and Mathematical Morphology*, Academic Press, New York, 1982.
8. Haralick, R. M., Sternberg, S. R., and Zhuang, X., Image analysis using mathematical morphology, *IEEE Trans. Pattern Analysis and Machine Intelligence*, 9, 532, 1988.
9. Shih, F. Y. and Mitchell, O. R., Threshold decomposition of grayscale morphology into binary morphology, *IEEE Trans. Pattern Analysis and Machine Intelligence*, 11, 31, 1989.
10. Dittmann, J. and Nack, F., Copyright-copywrong, *IEEE Multimedia*, 7, 14, 2000.
11. Guo, C., Ma, Q., and Zhang, L., Spatio-temporal saliency detection using phase spectrum of quaternion Fourier transform, in *Proc. IEEE Conf. Computer Vision and Pattern Recognition*, Anchorage, Alaska, 2008.
12. Hou, X. and Zhang, L., Saliency detection: A spectral residual approach, in *Proc. IEEE Conf. Computer Vision and Pattern Recognition*, Minneapolis, MN, 2007.
13. Huang, J., Shi, Y., and Shi, Y., Embedding image watermarks in DC components, *IEEE Trans. Circuits and Systems for Video Technology*, 10, 974, 2000.
14. Itti, L., Koch, C., and Niebur, E., A model of saliency-based visual attention for rapid scene analysis, *IEEE Trans. Pattern Analysis and Machine Intelligence*, 11, 1254, 1998.
15. Alattar, A. M., Reversible watermark using the difference expansion of a generalized integer transform, *IEEE Trans. Image Processing*, 13, 1147, 2004.
16. Al-Qershi, O. M. and Khoo, B. E., Two-dimensional difference expansion (2D-DE) scheme with a characteristics-based threshold, *Signal Processing*, 93, 154, 2013.
17. Tian, J., Reversible data embedding using a difference expansion, *IEEE Trans. and Systems for Video Technology*, 13, 890, 2003.
18. Chang, C. C., Tsai, P., and Lin, C. C., SVD-based digital image watermarking scheme, *Pattern Recognition Letters*, 26, 1577, 2005.
19. Wang, Y. R., Lin, W. H., and Yang, L., An intelligent watermarking method based on particle swarm optimization, *Expert Systems with Applications*, 38, 8024, 2011.
20. Qin, C., Chang, C. C., and Chiu, Y. P., A novel joint data-hiding and compression scheme based on SMVQ and image inpainting, *IEEE Trans. Image Processing*, 23, 969, 2014.
21. Qin, C. Chang, C. C., and Hsu, T. J., Reversible data hiding scheme based on exploiting modification direction with two steganographic images, *Multimedia Tools and Applications*, 74, 5861, 2015.

22. Qin, C. and Zhang, X., Effective reversible data hiding in encrypted image with privacy protection for image content, *J. Visual Communication and Image Representation*, 31, 154, 2015.

23. Kennedy, J., Particle swarm optimization, in *Encyclopedia of Machine Learning*, Springer, New York, 760, 2010.

24. Shih, F. Y. and Wu, Y. T., Robust watermarking and compression for medical images based on genetic algorithms, *Information Sciences*, 175, 200, 2005.

25. Ruderman, D. L. and Bialek, W., Statistics of natural images: Scaling in the woods, *Physical Review Letters*, 73, 814, 1994.

26. Srivastava, A., Lee, A. B., Simoncelli, E. P., and Zhu, S. C., On advances in statistical modeling of natural images, *J. Mathematical Imaging and Vision*, 18, 17, 2003.

27. Chadha, R. and Allison, D., Decomposing rectilinear figures into rectangles, Technical Report, Department of Computer Science, Virginia Polytechnic Institute and State University, 1988.

28. Wang, Z. and Bovik, A. C., A universal image quality index, *IEEE Signal Processing Letters*, 9, 81, 2002.

# 9 High-Capacity Watermarking

Since multimedia technologies have been becoming increasingly sophisticated in their rapidly growing Internet applications, data security, including copyright protection and data integrity detection, has raised tremendous concerns. One solution to achieving data security is digital watermarking technology, which embeds hidden information or secret data into a host image [1–3]. It serves as a suitable tool to identify the source, creator, owner, distributor, or authorized consumer of a document or image. It can also be used to detect whether a document or image has been illegally distributed or modified. This chapter is focused on high-capacity watermarking techniques, including robust watermarking and high-capacity, multiple-regions-of-interest watermarking for medical images.

## 9.1 ROBUST HIGH-CAPACITY DIGITAL WATERMARKING

There are two domain-based watermarking techniques: one in the spatial domain and the other in the frequency domain. In the spatial domain [4–8], we embed watermarks into a host image by changing the gray levels of certain pixels. The embedding capacity may be large, but the hidden information could be easily detected by means of computer analysis. In the frequency domain [9–17], we insert watermarks into the frequency coefficients of the image transformed by the *discrete Fourier transform* (DFT), *discrete cosine transform* (DCT), or *discrete wavelet transform* (DWT). The hidden information is in general difficult to detect, but we cannot embed a large volume of data in the frequency domain due to image distortion.

It is reasonable to think that frequency domain watermarking would be robust since embedded watermarks are spread out all over the spatial extent of an image [9]. If the watermarks were embedded into locations of large absolute values (the *significant coefficients*) of the transformed image, the watermarking technique would become more robust. Unfortunately, the transformed images in general contain only a few significant coefficients, so the watermarking capacity is limited.

### 9.1.1 WEAKNESS OF CURRENT ROBUST WATERMARKING

There are several approaches to achieving robust watermarking such that the watermarks are still detectable after applying distortion. However, the capacity of robust watermarking is usually limited due to its strategy. For example, the redundant embedding approach achieves robustness by embedding more than one copy of the same watermark into an image. However, the multiple copies reduce the size of the watermarks. That is, the greater the number of copies, the smaller the size of the watermarks. For significant-coefficients embedding, it is obvious that the capacity for watermarks depends on the number of significant coefficients. Unfortunately, the number of significant coefficients is quite limited in most real images due to local spatial similarity. For region-based embedding, we embed a bit of the watermark into a region of the image in order to spread the message all over the selected region. Hence, the capacity of watermarks is restricted by block size.

### 9.1.2  CONCEPT OF ROBUST WATERMARKING

Zhao et al. [10] presented a robust wavelet domain watermarking algorithm based on chaotic maps. They divide an image into a set of $8 \times 8$ blocks with labels. After the order is mixed by a chaotic map, the first 256 blocks are selected for embedding. Miller et al. [11] proposed a *robust high-capacity* (RHC) watermarking algorithm by informed coding and embedding. However, they can embed only 1380 bits of information in an image of size $240 \times 368$—that is, a capacity of 0.015625 bits per pixel (BPP).

There are two criteria to be considered when we develop a RHC watermarking technique. First, the strategy for embedding and extracting watermarks ensures the robustness. Second, the strategy for enlarging the capacity does not affect the robustness. Frequency domain watermarking is robust because the embedded messages are spread all over the spatial extent of the image [9]. Moreover, if the messages are embedded into the significant coefficients, the watermarking technique will be even more robust. Therefore, the significant-coefficients embedding approach is adopted as the base of the RHC technique to satisfy the first criterion. The remaining problem is to enlarge the capacity without degrading the robustness.

### 9.1.3  ENLARGEMENT OF SIGNIFICANT COEFFICIENTS

During spatial frequency transformation, the low frequencies in the transformed domain reflect the smooth areas of an image, and the high frequencies reflect areas with high intensity changes, such as edges and noise. Therefore, due to local spatial similarity, the significant coefficients of a transformed image are limited. Unless an image contains a lot of noise, we cannot obtain many significant coefficients through any of the usual transformation approaches (DCT, DFT, or DWT).

#### 9.1.3.1  Breaking Local Spatial Similarity

A noisy image gives us a larger capacity. If we could rearrange pixel positions such that the rearranged image contains lots of noise, the number of significant coefficients in the image would be increased dramatically. Therefore, we adopt a chaotic map to relocate pixels.

Figure 9.1 illustrates an example of increasing the number of significant coefficients by rearranging pixel locations. Figure 9.1a shows an image whose gray values are displayed in Figure 9.1b. Figure 9.1d shows the relocated image, with gray values displayed in Figure 9.1e. Figures 9.1c,f are obtained by applying DCT on Figures 9.1b and 9.1e, respectively. If the threshold is set to be 70, we obtain 14 significant coefficients in Figure 9.1f, but only 2 significant coefficients in Figure 9.1c.

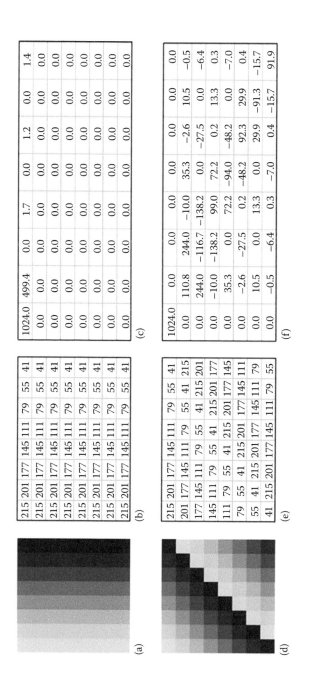

**FIGURE 9.1** An example of increasing the number of significant coefficients: (a,b) an image and its gray values, (c) the DCT coefficients, (d,e) the relocated image and its gray values, (f) the DCT coefficients.

### 9.1.3.2    Block-Based Chaotic Map

In order to enlarge the watermark capacity, we adopt a chaotic map to break the local spatial similarity of an image so it can generate more significant coefficients in the transformed domain. Since the traditional chaotic map is pixel based [6,10,16] and not suited to generating the reference register, we have developed a new block-based chaotic map to break the local spatial similarity. In other words, the block-based chaotic map for relocating pixels is based on a *block* unit (i.e., a set of connected pixels) instead of a *pixel* unit.

Figures 9.2a,b show a diagram and its relocated result based on a block of size $2 \times 2$. Figures 9.2c,d show a diagram and its relocated result based on a block of size $2 \times 4$. In general, the bigger the block size, the greater the local similarity. We apply a block of size $2 \times 2$ and $l = 2$ in Figure 9.3, where (a) is the Lena image, (b) is its relocated image, and (c) is the resulting image in the next iteration.

### 9.1.4    Determination of Embedding Locations

#### 9.1.4.1    Intersection-Based Pixel Collection

*Intersection-based pixel collection* (IBPC) intends to label an image using two symbols alternatively and then collect the two subimages with the same symbol. Figures 9.4a–c show three different approaches to collecting pixels with same label horizontally, vertically, and diagonally, respectively. The two subimages formed are shown in Figures 9.4d,e. Note that the pair of images obtained have local spatial similarity, even after transformation or attacks. After the experiments, diagonal IBPC shows better similarity in our RHC watermarking algorithm. Therefore, we generate a pair of $8 \times 8$ coefficients from the $16 \times 8$ image by using the IBPC approach

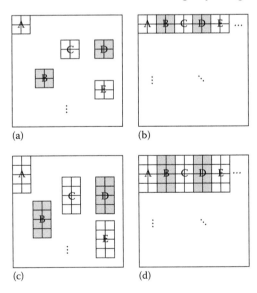

(a)                    (b)

(c)                    (d)

**FIGURE 9.2**    An example of block-based relocation: (a,b) a diagram and its relocated result based on the $2 \times 2$ block, (c,d) a diagram and its relocated result based on the $2 \times 4$ block.

(a)                              (b)                              (c)

**FIGURE 9.3**   An example of performing a block-based chaotic map, (a) a Lena image, (b) the relocated image, (c) the continuously relocated image.

followed by the DCT. Afterward, one is used as the reference register for indicating significant DCT coefficients, and the other is used as the container for embedding watermarks.

### 9.1.4.2   Reference Register and Container

An example of demonstrating the similarity of two subimages obtained by IBPC is shown in Figure 9.5. Figure 9.5a shows a Lena image, from which a small block of pixels have been cropped in Figure 9.5b. Figures 9.5c,d show the pair of subimages obtained from Figure 9.5b by IBPC. Figures 9.5e,f are their respective DCT images. Figures 9.5g,h show the results after dividing Figures 9.5e and 9.5f, respectively, by the quantization table with a *quality factor* (QF) of 50. Note that the QF will be used in Equation 9.1, and an example of the quantization table in JPEG compression is given in Figure 9.6. We observe that Figures 9.5g,h are similar. Therefore, either one can be used as the reference register to indicate the significant coefficients for embedding watermarks. The significant coefficients have values above a predefined threshold $RR_{Th}$.

Let $QTable(i,j)$ denote the quantization table, where $0 \leq i,j \leq 7$. The new quantization table $NewTable(i,j)$ is obtained by

$$NewTable(i, j) = \begin{cases} QTable(i, j) \times \dfrac{50}{QF} & \text{if } QF < 50 \\ QTable(i, j) \times (2 - 0.02 \times QF) & \text{otherwise} \end{cases}. \qquad (9.1)$$

### 9.1.5   RHC WATERMARKING ALGORITHM

Our RHC watermarking algorithm contains two main components: a *block-based chaotic map* (BBCM) and a *reference register* (RR).

### 9.1.5.1   Embedding Procedure

In order to explain our watermark-embedding procedures clearly, we must first introduce the following symbols:

$H$: The gray-level host image of size $n \times m$.

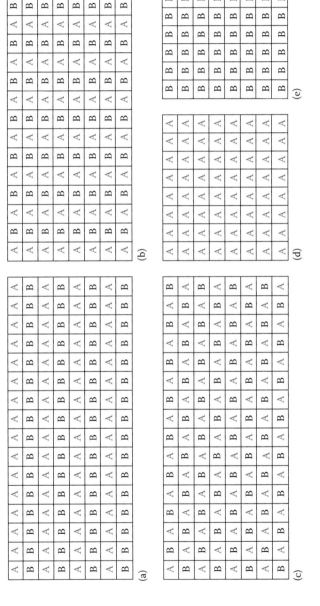

**FIGURE 9.4**    An example of IBPC.

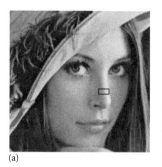

(a)

| | | | | | | | | | | | | | | | |
|---|---|---|---|---|---|---|---|---|---|---|---|---|---|---|---|
| 129 | 132 | 133 | 131 | 137 | 151 | 152 | 161 | 176 | 198 | 207 | 220 | 223 | 224 | 221 | 212 |
| 129 | 133 | 132 | 137 | 134 | 143 | 149 | 164 | 178 | 193 | 204 | 218 | 221 | 222 | 220 | 216 |
| 130 | 133 | 133 | 133 | 137 | 144 | 151 | 167 | 174 | 187 | 210 | 215 | 219 | 221 | 219 | 216 |
| 131 | 128 | 134 | 130 | 141 | 147 | 152 | 160 | 170 | 181 | 203 | 215 | 220 | 220 | 222 | 216 |
| 127 | 128 | 131 | 136 | 141 | 144 | 150 | 159 | 168 | 186 | 195 | 208 | 220 | 224 | 222 | 217 |
| 121 | 130 | 130 | 131 | 137 | 144 | 154 | 157 | 166 | 177 | 186 | 208 | 215 | 223 | 223 | 219 |
| 125 | 131 | 129 | 132 | 137 | 145 | 147 | 154 | 161 | 171 | 184 | 207 | 216 | 222 | 223 | 216 |
| 124 | 124 | 132 | 137 | 136 | 142 | 145 | 150 | 163 | 171 | 187 | 203 | 216 | 220 | 223 | 218 |

(b)

| | | | | | | | |
|---|---|---|---|---|---|---|---|
| 129 | 133 | 137 | 152 | 176 | 207 | 223 | 221 |
| 133 | 137 | 143 | 164 | 193 | 218 | 222 | 216 |
| 130 | 133 | 137 | 151 | 174 | 210 | 219 | 219 |
| 128 | 130 | 147 | 160 | 181 | 215 | 220 | 216 |
| 127 | 131 | 141 | 150 | 168 | 195 | 220 | 222 |
| 130 | 131 | 144 | 157 | 177 | 208 | 223 | 219 |
| 125 | 129 | 137 | 147 | 161 | 184 | 216 | 223 |
| 124 | 137 | 142 | 150 | 171 | 203 | 220 | 218 |

(c)

| | | | | | | | |
|---|---|---|---|---|---|---|---|
| 132 | 131 | 151 | 161 | 198 | 220 | 224 | 212 |
| 129 | 132 | 134 | 149 | 178 | 204 | 221 | 220 |
| 133 | 133 | 144 | 167 | 187 | 215 | 221 | 216 |
| 131 | 134 | 141 | 152 | 170 | 203 | 220 | 222 |
| 128 | 136 | 144 | 159 | 186 | 208 | 224 | 217 |
| 121 | 130 | 137 | 154 | 166 | 186 | 215 | 223 |
| 131 | 132 | 145 | 154 | 171 | 207 | 222 | 216 |
| 124 | 132 | 136 | 145 | 163 | 187 | 216 | 223 |

(d)

| | | | | | | | |
|---|---|---|---|---|---|---|---|
| 1375.5 | −285.6 | −28.0 | 22.7 | −22.5 | 7.6 | 0.9 | −1.9 |
| 17.5 | −3.0 | −11.5 | 15.6 | 1.9 | −2.9 | 2.1 | −2.9 |
| −2.3 | 0.0 | 3.7 | 0.2 | −0.7 | −3.7 | −5.3 | −0.5 |
| −3.1 | −1.4 | 5.5 | −4.8 | 3.7 | 3.9 | −2.3 | 3.1 |
| −2.0 | −1.4 | 0.8 | −2.6 | −4.5 | −0.1 | 0.7 | 0.4 |
| −11.0 | −1.3 | 6.4 | −5.0 | 1.3 | 2.8 | 1.7 | −0.2 |
| 1.7 | −2.8 | 1.7 | 5.1 | −4.9 | −0.3 | 2.3 | −2.5 |
| −19.6 | −0.3 | 18.4 | −9.3 | 1.9 | −0.2 | −4.1 | 1.0 |

(e)

| | | | | | | | |
|---|---|---|---|---|---|---|---|
| 1377.9 | 283.2 | 23.4 | 22.2 | −18.4 | 7.8 | −2.0 | 1.9 |
| 23.5 | −2.6 | −18.8 | 18.6 | 1.0 | −2.7 | 4.1 | 2.6 |
| −3.5 | −0.6 | −0.1 | 2.9 | −1.4 | −1.1 | 2.2 | 4.7 |
| 3.8 | −0.9 | −8.9 | −0.1 | −0.5 | −0.5 | −0.6 | 4.9 |
| 4.6 | 0.3 | −1.8 | 1.4 | −2.6 | −3.0 | 0.6 | 4.7 |
| 10.7 | 3.2 | −6.0 | 4.3 | −4.3 | 5.0 | 8.1 | −1.2 |
| −0.9 | 1.3 | −6.3 | −6.4 | 6.1 | 0.4 | −1.7 | −0.4 |
| 22.0 | 1.4 | 18.8 | 12.0 | −2.8 | 1.6 | 2.9 | 1.2 |

(f)

| | | | | | | | |
|---|---|---|---|---|---|---|---|
| 83.4 | −24.8 | 2.7 | 1.4 | −0.9 | 0.2 | 0.0 | 0.0 |
| 1.4 | −0.2 | −0.8 | 0.8 | 0.1 | 0.0 | 0.0 | 0.0 |
| −0.2 | 0.0 | 0.2 | 0.0 | 0.0 | −0.1 | −0.1 | −0.1 |
| −0.2 | −0.1 | 0.2 | −0.2 | 0.1 | 0.0 | 0.0 | 0.0 |
| −0.1 | −0.1 | 0.0 | 0.0 | −0.1 | 0.0 | 0.0 | 0.0 |
| −0.4 | 0.0 | 0.1 | −0.1 | 0.0 | 0.0 | 0.0 | 0.0 |
| 0.0 | 0.0 | 0.0 | 0.1 | 0.0 | 0.0 | 0.0 | 0.0 |
| −0.3 | 0.0 | 0.2 | −0.1 | 0.0 | 0.0 | 0.0 | 0.0 |

(g)

| | | | | | | | |
|---|---|---|---|---|---|---|---|
| 83.5 | 24.6 | 2.2 | 1.3 | −0.8 | 0.2 | 0.0 | 0.0 |
| 1.9 | −0.2 | −1.3 | 1.0 | 0.0 | 0.0 | 0.1 | 0.0 |
| −0.2 | 0.0 | 0.0 | 0.1 | 0.0 | 0.0 | 0.0 | 0.1 |
| 0.3 | −0.1 | −0.4 | 0.0 | 0.0 | 0.0 | 0.0 | 0.1 |
| 0.2 | 0.0 | 0.0 | 0.0 | 0.0 | 0.0 | 0.0 | 0.1 |
| 0.4 | 0.1 | −0.1 | 0.1 | −0.1 | 0.0 | 0.1 | 0.0 |
| 0.0 | 0.0 | −0.1 | −0.1 | 0.0 | 0.0 | 0.0 | 0.0 |
| 0.3 | 0.0 | 0.2 | 0.1 | 0.0 | 0.0 | 0.0 | 0.0 |

(h)

**FIGURE 9.5** An example of illustrating the similarity of two subimages obtained by IBPC: (a) a Lena image, (b) the extracted $16 \times 8$ image, (c,d) a pair of extracted $8 \times 8$ subimages, (e,f) the $8 \times 8$ DCT coefficients, (g,h) the quantized images.

| | | | | | | | |
|---|---|---|---|---|---|---|---|
| 16 | 11 | 10 | 16 | 24 | 40 | 51 | 61 |
| 12 | 12 | 14 | 19 | 26 | 58 | 60 | 55 |
| 14 | 13 | 16 | 24 | 40 | 57 | 69 | 56 |
| 14 | 17 | 22 | 29 | 51 | 87 | 80 | 62 |
| 18 | 22 | 37 | 56 | 68 | 109 | 103 | 77 |
| 24 | 35 | 55 | 64 | 81 | 104 | 113 | 92 |
| 49 | 64 | 78 | 87 | 103 | 121 | 120 | 101 |
| 72 | 92 | 95 | 98 | 112 | 100 | 103 | 99 |

**FIGURE 9.6** A quantization table for JPEG compression.

$H_{(i,j)}^{16\times8}$ : The $(i,j)$th subimage of size $16\times8$ of $H$, where $1 \le i \le [n/16]$ and $1 \le j \le [m/8]$.

$HA_{(i,j)}^{8\times8}$ and $HB_{(i,j)}^{8\times8}$ : A pair of images obtained from $H_{(i,j)}^{16\times8}$ by IBPC.

$DA_{(i,j)}^{8\times8}$ and $DB_{(i,j)}^{8\times8}$ : The images transformed by applying the DCT on $HA_{(i,j)}^{8\times8}$ and $HB_{(i,j)}^{8\times8}$, respectively.

$QA_{(i,j)}^{8\times8}$ and $QB_{(i,j)}^{8\times8}$ : The images that result from dividing $DA_{(i,j)}^{8\times8}$ and $DB_{(i,j)}^{8\times8}$ by the quantization table, respectively.

$E_{(i,j)}^{8\times8}$ : The watermarked image of $DA_{(i,j)}^{8\times8}$ using $QA_{(i,j)}^{8\times8}$ and $QB_{(i,j)}^{8\times8}$ .

$M_{(i,j)}^{8\times8}$ : The image after multiplying $E_{(i,j)}^{8\times8}$ by the quantization table.

$I_{(i,j)}^{8\times8}$ : The image obtained by applying the *inverse discrete cosine transform* (IDCT) to $M_{(i,j)}^{8\times8}$.

$C_{(i,j)}^{16\times8}$ : The watermarked subimage obtained by combining $I_{(i,j)}^{8\times8}$ and $HB_{(i,j)}^{8\times8}$ using IBPC.

$O$: The output watermarked image obtained by collecting all the subimages of $C_{(i,j)}^{16\times8}$ .

We present the overall embedding procedure as follows and the flowchart in Figure 9.7.

*The embedding procedures of our RHC watermarking algorithm*:

1. Divide an input image $H$ into a set of subimages, $H_{(i,j)}^{16\times8}$ , of size $16\times9$.
2. Build $HA_{(i,j)}^{8\times8}$ and $HB_{(i,j)}^{8\times8}$ from each subimage $H_{(i,j)}^{16\times8}$ by IBPC.
3. Obtain $DA_{(i,j)}^{8\times8}$ and $DB_{(i,j)}^{8\times8}$ from $HA_{(i,j)}^{8\times8}$ and $HB_{(i,j)}^{8\times8}$ with the DCT, respectively.
4. Obtain $QA_{(i,j)}^{8\times8}$ and $QB_{(i,j)}^{8\times8}$ by dividing $DA_{(i,j)}^{8\times8}$ and $DB_{(i,j)}^{8\times8}$ by the JPEG quantization table, respectively.
5. Determine the proper positions of the significant coefficients in $QB_{(i,j)}^{8\times8}$ and embed the watermarks into the corresponding positions in $QA_{(i,j)}^{8\times8}$ to obtain $E_{(i,j)}^{8\times8}$. A detailed embedding strategy will be described in Section 9.1.5.3.
6. Obtain $M_{(i,j)}^{8\times8}$ by multiplying $E_{(i,j)}^{8\times8}$ by the JPEG quantization table.

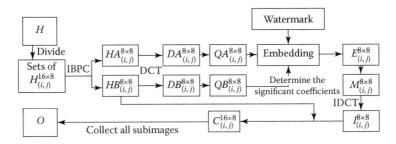

**FIGURE 9.7**    The embedding procedure of our RHC watermarking algorithm.

7. Obtain $I_{(i,j)}^{8\times8}$ by applying the IDCT to $M_{(i,j)}^{8\times8}$.

8. Reconstruct $C_{(i,j)}^{16\times8}$ by combining $I_{(i,j)}^{8\times8}$ and $HB_{(i,j)}^{8\times8}$.

9. Obtain the output watermarked image $O$ by collecting $C_{(i,j)}^{16\times8}$.

### 9.1.5.2 Extraction Procedure

After receiving the watermarked image $O$, we intend to extract the watermark information. The following symbols are introduced in order to explain the watermark extraction procedures.

$O_{(i,j)}^{16\times8}$: The $(i,j)$th subimage of size $16\times8$ of $O$.

$OA_{(i,j)}^{8\times8}$ and $OB_{(i,j)}^{8\times8}$: A pair of extracted images from $O_{(i,j)}^{16\times8}$ using IBPC.

$TA_{(i,j)}^{8\times8}$ and $TB_{(i,j)}^{8\times8}$: The image transformed by applying the DCT to $OA_{(i,j)}^{8\times8}$ and $OB_{(i,j)}^{8\times8}$, respectively.

$RA_{(i,j)}^{8\times8}$ and $RB_{(i,j)}^{8\times8}$: The images that result from dividing $TA_{(i,j)}^{8\times8}$ and $TB_{(i,j)}^{8\times8}$ by the quantization table, respectively.

We present the watermark extraction procedure as follows and the flowchart in Figure 9.8.

*The extraction procedures of our RHC watermarking algorithm*:

1. Divide the watermarked image $O$ into a set of subimages $O_{(i,j)}^{16\times8}$ of size $16\times9$.

2. Build $OA_{(i,j)}^{8\times8}$ and $OB_{(i,j)}^{8\times8}$ from each subimage $O_{(i,j)}^{16\times8}$ by IBPC.

3. Obtain $TA_{(i,j)}^{8\times8}$ and $TB_{(i,j)}^{8\times8}$ from $OA_{(i,j)}^{8\times8}$ and $OB_{(i,j)}^{8\times8}$ by the DCT, respectively.

4. Obtain $RA_{(i,j)}^{8\times8}$ and $RB_{(i,j)}^{8\times8}$ by dividing $TA_{(i,j)}^{8\times8}$ and $TB_{(i,j)}^{8\times8}$ by the JPEG quantization table, respectively.

5. Determine the proper positions of the significant coefficients in $RB_{(i,j)}^{8\times8}$ and extract the subwatermark from the corresponding positions in $RA_{(i,j)}^{8\times8}$. A detailed extraction strategy will be described in Section 9.1.5.3.

6. Collect all the subwatermarks to obtain the watermark.

### 9.1.5.3 Embedding and Extraction Strategies

In this section, we will describe the embedding and extraction strategies of our RHC watermarking algorithm. For embedding, each pair of $QA_{(i,j)}^{8\times8}$ (container) and $QB_{(i,j)}^{8\times8}$ (reference register) is obtained first. After the significant coefficients are determined

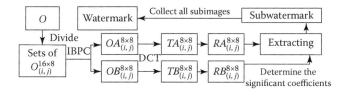

**FIGURE 9.8** The extraction procedures of our RHC watermarking algorithm.

by the reference register, the watermarks are embedded into the corresponding positions of the cover coefficients by adding the values in Equation 9.2. Let $V_{(k,l)}^C$ and $V_{(k,l)}^R$ denote the values of $QA_{(i,j)}^{8\times8}$ and $QB_{(i,j)}^{8\times8}$, respectively. Let $S_{(k,l)}^C$ be the result after embedding the message.

$$S_{(k,l)}^C = \begin{cases} V_{(k,l)}^C + \alpha V_{(k,l)}^R & \text{if } V_{(k,l)}^R \geq RR_{Th} \\ V_{(k,l)}^C & \text{otherwise} \end{cases}, \qquad (9.2)$$

where $0 \leq k,l \leq 7$ and $\alpha > 0$. Note that the bigger $\alpha$ is, the higher the robustness. The $8 \times 8$ $S_{(k,l)}^C$ s are collected to form the corresponding $E_{(i,j)}^{8\times8}$. The embedding strategy is as follows.

*Watermark-embedding strategy*:

1. Determine the embedding location by checking the significant coefficients of $QB_{(i,j)}^{8\times8}$.
2. If the embedding message is 1 and $V_{(k,l)}^R \geq RR_{Th}$, we obtain $S_{(k,l)}^C = V_{(k,l)}^C + \alpha V_{(k,l)}^R$; otherwise, we set $S_{(k,l)}^C = V_{(k,l)}^C$.

For the extraction strategy, each pair of $RA_{(i,j)}^{8\times8}$ (container) and $RB_{(i,j)}^{8\times8}$ (reference register) is obtained first. After the embedded positions of the reference register are determined, the watermark can be extracted by Equation 9.3, and the watermarked coefficients can be used to construct the original ones with Equation 9.4. Let $W_{(k,l)}^C$ and $W_{(k,l)}^R$ denote the values of $RA_{(i,j)}^{8\times8}$ and $RB_{(i,j)}^{8\times8}$, respectively. Let $F_{(k,l)}^C$ be the result after the additional amount is removed by Equation 9.4 and $w$ be the embedded message.

$$w = \begin{cases} 1 & \text{if } (W_{(k,l)}^C - W_{(k,l)}^R) \geq \dfrac{\alpha}{2} V_{(k,l)}^R \\ 0 & \text{elsewhere} \end{cases}. \qquad (9.3)$$

$$F_{(k,l)}^C = \begin{cases} W_{(k,l)}^C - \alpha W_{(k,l)}^R & \text{if } w = 1 \\ W_{(k,l)}^C & \text{otherwise} \end{cases}. \qquad (9.4)$$

The embedding strategy of our RHC watermarking is as follows.
*Watermark extraction strategy*:

1. Determine the extraction location by checking the significant coefficients of $RB_{(i,j)}^{8\times8}$.
2. Obtain the embedded message using Equation 9.2.
3. If the embedded message is 1, we calculate $F_{(k,l)}^C = W_{(k,l)}^C - \alpha W_{(k,l)}^R$; otherwise, we set $F_{(k,l)}^C = W_{(k,l)}^C$.

### 9.1.6 Experimental Results

In this section, we will show an example to illustrate that our algorithm can enlarge the watermarking capacity by using the BBCM and IBPC approaches. We will provide experimental results for 200 images and comparisons with the *iciens* algorithm by Miller et al. [11].

#### 9.1.6.1 Capacity Enlargement

We give an example of the embedding strategy in Figure 9.9. Figure 9.9a is the relocated image of Figure 9.3a after performing seven iterations of relocation using the BBCM in Equation 9.1, where the block size is $2 \times 2$ and $l = 2$. After obtaining Figure 9.9b from the small box in Figure 9.9a, we generate Figures 9.9c,d by IBPC. Figures 9.9e,f are respectively obtained from Figures 9.9c,d by the DCT followed by

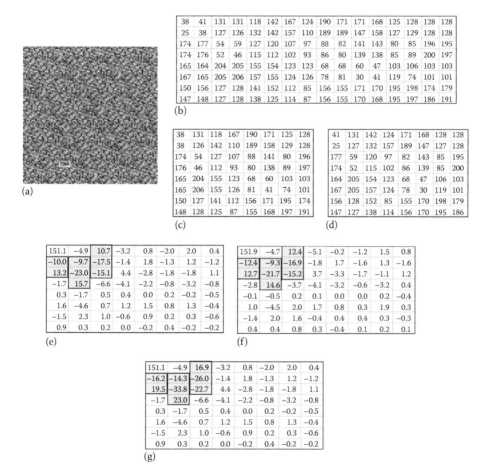

**FIGURE 9.9** An example of the embedding strategy: (a) a relocated image, (b) a $16 \times 8$ subimage, (c,d) a pair of extracted results using IBPC, (e) a container, (f) a reference register, (g) the watermarked result.

a division of the quantization table. Note that Figure 9.9e is considered the container for watermark embedding, and Figure 9.9f is the reference register for indicating the embedding positions, as colored in gray. Since the number of embedding positions is eight, we set the watermark to be 11111111. Figure 9.9g shows the results after embedding watermarks into these calculated positions with Equation 9.2, where $\alpha = 0.5$ and $RR_{Th} = 9$.

Figure 9.10 shows an example of the extraction strategy. Figure 9.10a is the watermarked subimage, and Figures 9.10b,c are extracted from Figure 9.10a by IBPC. Note that since Figure 9.10c is exactly the same as Figure 9.9d, the reference register is the same as Figure 9.9f. Therefore, after obtaining Figure 9.10d from 9.10b by the DCT, we can extract the watermark 11111111 from the coefficients and generate Figure 9.10e. Figure 9.10f is the reconstructed result. We observe that Figure 9.10f is exactly the same as the original image in Figure 9.9b.

(a)

| 11 | 41 | 116 | 131 | 121 | 142 | 184 | 124 | 211 | 171 | 184 | 168 | 126 | 128 | 119 | 128 |
|---|---|---|---|---|---|---|---|---|---|---|---|---|---|---|---|
| 25 | 12 | 127 | 108 | 132 | 136 | 157 | 115 | 189 | 199 | 147 | 168 | 127 | 134 | 128 | 129 |
| 160 | 177 | 40 | 59 | 114 | 120 | 97 | 97 | 82 | 82 | 140 | 143 | 84 | 85 | 203 | 195 |
| 174 | 188 | 52 | 49 | 115 | 102 | 102 | 73 | 86 | 58 | 139 | 122 | 85 | 82 | 200 | 197 |
| 200 | 164 | 225 | 205 | 155 | 154 | 104 | 123 | 40 | 68 | 35 | 47 | 86 | 106 | 93 | 103 |
| 167 | 203 | 205 | 232 | 157 | 162 | 124 | 115 | 78 | 61 | 30 | 23 | 119 | 63 | 101 | 95 |
| 169 | 156 | 140 | 128 | 145 | 152 | 110 | 85 | 153 | 155 | 173 | 170 | 205 | 198 | 190 | 179 |
| 147 | 146 | 127 | 126 | 138 | 124 | 114 | 90 | 156 | 166 | 170 | 189 | 195 | 227 | 186 | 227 |

(b)

| 11 | 116 | 121 | 184 | 211 | 184 | 126 | 119 |
|---|---|---|---|---|---|---|---|
| 12 | 108 | 136 | 115 | 199 | 168 | 134 | 129 |
| 160 | 40 | 114 | 97 | 82 | 140 | 84 | 203 |
| 188 | 49 | 102 | 73 | 58 | 122 | 82 | 197 |
| 200 | 225 | 155 | 104 | 40 | 35 | 86 | 93 |
| 203 | 232 | 162 | 115 | 61 | 23 | 63 | 95 |
| 169 | 140 | 145 | 110 | 153 | 173 | 205 | 190 |
| 146 | 126 | 124 | 90 | 166 | 189 | 227 | 227 |

(c)

| 41 | 131 | 142 | 124 | 171 | 168 | 128 | 128 |
|---|---|---|---|---|---|---|---|
| 25 | 127 | 132 | 157 | 189 | 147 | 127 | 128 |
| 177 | 59 | 120 | 97 | 82 | 143 | 85 | 195 |
| 174 | 52 | 115 | 102 | 86 | 139 | 85 | 200 |
| 164 | 205 | 154 | 123 | 68 | 47 | 106 | 103 |
| 167 | 205 | 157 | 124 | 78 | 30 | 119 | 101 |
| 156 | 128 | 152 | 85 | 155 | 170 | 198 | 179 |
| 147 | 127 | 138 | 114 | 156 | 170 | 195 | 186 |

(d)

| 151.0 | −5.0 | 16.8 | −3.1 | 0.8 | −2.0 | 2.0 | 0.4 |
|---|---|---|---|---|---|---|---|
| −16.2 | −14.3 | −26.1 | −1.3 | 1.8 | −1.3 | 1.2 | −1.2 |
| 19.5 | −33.9 | −22.7 | 4.4 | −2.7 | −1.8 | −1.8 | 1.1 |
| −1.7 | 23.0 | −6.6 | −4.1 | −2.3 | −0.8 | −3.2 | −0.8 |
| 0.3 | −1.7 | 0.5 | 0.4 | 0.1 | 0.2 | −0.2 | −0.5 |
| 1.6 | −4.6 | 0.7 | 1.2 | 1.5 | 0.8 | 1.3 | −0.3 |
| −1.5 | 2.3 | 1.0 | −0.6 | 0.9 | 0.2 | 0.3 | −0.6 |
| 0.9 | 0.2 | 0.2 | 0.0 | −0.2 | 0.4 | −0.2 | −0.2 |

(e)

| 151.0 | −5.0 | 10.6 | −3.1 | 0.8 | −2.0 | 2.0 | 0.4 |
|---|---|---|---|---|---|---|---|
| −10.0 | −9.7 | −17.6 | −1.3 | 1.8 | −1.3 | 1.2 | −1.2 |
| 13.2 | −23.0 | −15.1 | 4.4 | −2.7 | −1.8 | −1.8 | 1.1 |
| −1.7 | 15.7 | −6.6 | −4.1 | −2.3 | −0.8 | −3.2 | −0.8 |
| 0.3 | −1.7 | 0.5 | 0.4 | 0.1 | 0.2 | −0.2 | −0.5 |
| 1.6 | −4.6 | 0.7 | 1.2 | 1.5 | 0.8 | 1.3 | −0.3 |
| −1.5 | 2.3 | 1.0 | −0.6 | 0.9 | 0.2 | 0.3 | −0.6 |
| 0.9 | 0.2 | 0.2 | 0.0 | −0.2 | 0.4 | −0.2 | −0.2 |

(f)

| 38 | 41 | 131 | 131 | 118 | 142 | 167 | 124 | 190 | 171 | 171 | 168 | 125 | 128 | 128 | 128 |
|---|---|---|---|---|---|---|---|---|---|---|---|---|---|---|---|
| 25 | 38 | 127 | 126 | 132 | 142 | 157 | 110 | 189 | 189 | 147 | 158 | 127 | 129 | 128 | 128 |
| 174 | 177 | 54 | 59 | 127 | 120 | 107 | 97 | 88 | 82 | 141 | 143 | 80 | 85 | 196 | 195 |
| 174 | 176 | 52 | 46 | 115 | 112 | 102 | 93 | 86 | 80 | 139 | 138 | 85 | 89 | 200 | 197 |
| 165 | 164 | 204 | 205 | 155 | 154 | 123 | 123 | 68 | 68 | 60 | 47 | 103 | 106 | 103 | 103 |
| 167 | 165 | 205 | 206 | 157 | 155 | 124 | 126 | 78 | 81 | 30 | 41 | 119 | 74 | 101 | 101 |
| 150 | 156 | 127 | 128 | 141 | 152 | 112 | 85 | 156 | 155 | 171 | 170 | 195 | 198 | 174 | 179 |
| 147 | 148 | 127 | 128 | 138 | 125 | 114 | 87 | 156 | 155 | 170 | 168 | 195 | 197 | 186 | 191 |

**FIGURE 9.10**  An example of the extraction strategy: (a) a $16 \times 8$ watermarked image, (b,c) a pair of extracted $8 \times 8$ results by IBPC, (d) an $8 \times 8$ watermarked image by the DCT, (e) the result after the embedded data are removed, (f) a reconstructed $16 \times 8$ image.

### 9.1.6.2 Robust Experiments

Miller et al. [11] present an RHC watermarking technique that applies informed coding and embedding. They conclude that the watermarking scheme is robust if the *message error rate* (MER) is lower than 20%. In order to evaluate the robustness of our watermarking algorithm, we tested it on 200 images and attacked the water-marked images with the JPEG compression, Gaussian noise, and low-pass filtering. The resulting MERs from these attacked images are provided.

Figure 9.11 shows the robustness experiment under JPEG compression. Figure 9.11a is the original Lena image. After continuously performing the BBCM nine times, we obtain Figure 9.11b, in which we embed watermarks by using the following parameters: a block size of $2 \times 2$, $l = 2$, $\alpha = 0.2$, and $RR_{Th} = 1$. Figure 9.11c is the watermarked image obtained by continuously performing the BBCM eight times to Figure 9.11b. Figure 9.11d is the attacked image, given by applying JPEG compression to Figure 9.11c using a QF of 20. Figure 9.11e is the reconstructed image after the embedded watermarks are removed. Note that our RHC watermark-ing is not only robust due to its MER of 0.19, which is less than 20%, but it also has a high capacity of 0.116 BPP, which is much larger than the 0.015625 BPP in Miller et al. [11].

Figures 9.12 and 9.13 show the experimental results after the low-pass filter and Gaussian noise attacks, respectively. The parameters used are the same as in the

**FIGURE 9.11**  An example of robustness experiment using JPEG compression: (a) a $202 \times 202$ Lena image, (b) the relocated image, (c) the watermarked image, (d) the image with JPEG compression, (e) the reconstructed image.

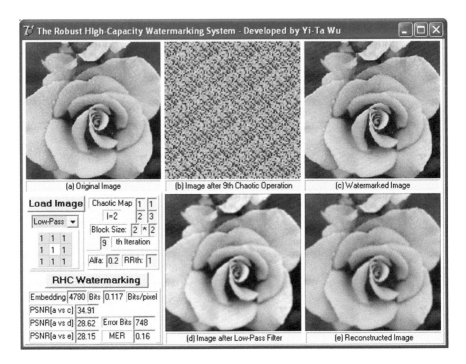

**FIGURE 9.12** An example robustness experiment using low-pass filter: (a) a $202 \times 202$ image, (b) the relocated image, (c) the watermarked image, (d) the image attacked by a low-pass filter, (e) the reconstructed image.

previous example. In Figure 9.12, a $3 \times 3$ low-pass filter with all 1s is utilized as the attacker. The capacity and the MER are 0.117 and 16%, respectively. In Figure 9.13, Gaussian noise with a *standard deviation* (SD) of 500 and a mean of 0 is added to the watermarked image. The capacity and the MER are 0.124 and 23%, respectively. It is obvious that our RHC watermarking algorithm not only achieves the robustness required but also maintains a high capacity for watermarks.

### 9.1.6.3  Performance Comparisons

In this section, we compare our RHC watermarking algorithm with the iciens algorithm by Miller et al. [11]. Figure 9.14 shows the effect under JPEG compression with different QFs. For simplicity, we use the symbol * to represent our RHC and ▲ to represent iciens. Our algorithm has a slightly higher MER when the QF is higher than 50, but has a significantly lower MER when the QF is lower than 40.

Figure 9.15 shows the effect under Gaussian noise with a mean of 0 and different standard deviations. Our RHC algorithm outperforms the iciens algorithm in terms of a low MER. Note that the parameters are a block size of $2 \times 2$, $l = 2$, $\alpha = 0.2$, and $RR_{Th} = 1$. For watermarking capacity, the average capacity of our RHC algorithm is 0.1238 BPP, which is much larger than the 0.015625 BPP obtained by Miller's iciens algorithm.

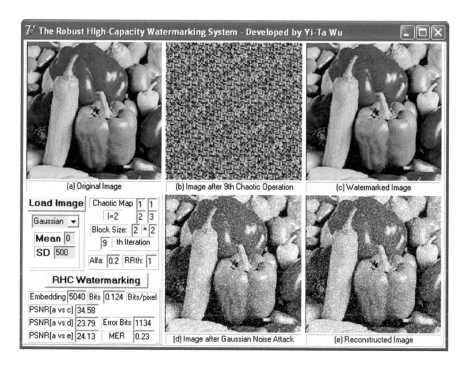

**FIGURE 9.13**   An example robustness experiment using Gaussian noise: (a) a $202 \times 202$ image, (b) the relocated image, (c) the watermarked image, (d) the image attacked by Gaussian noise, (e) the reconstructed image.

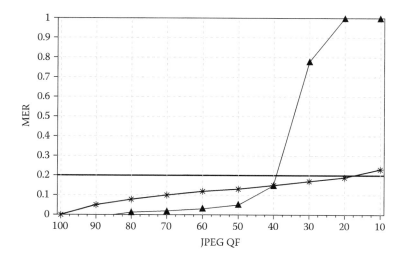

**FIGURE 9.14**   Robustness vs. JPEG compression.

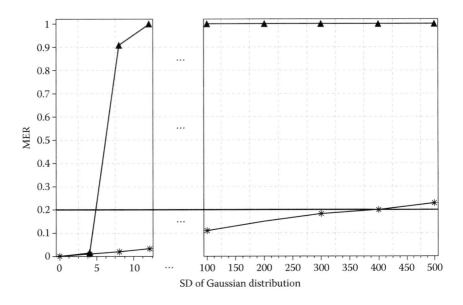

**FIGURE 9.15**    Robustness vs. Gaussian noise.

---

**TABLE 9.1**
**Effect of $RR_{Th}$**

| | | | JPEG (QF = 20) | | Gaussian noise (SD = 500, mean = 0) | |
|---|---|---|---|---|---|---|
| $RR_{Th}$ | Capacity(bits) | Capacity(BPP) | Error Bits | MER | Error Bits | MER |
| 10 | 810 | 0.0199 | 104 | 0.12 | 140 | 0.17 |
| 9 | 924 | 0.0226 | 120 | 0.13 | 170 | 0.18 |
| 8 | 1061 | 0.0260 | 139 | 0.13 | 195 | 0.18 |
| 7 | 1228 | 0.0301 | 163 | 0.13 | 228 | 0.19 |
| 6 | 1439 | 0.0353 | 197 | 0.14 | 271 | 0.19 |
| 5 | 1715 | 0.0420 | 243 | 0.14 | 329 | 0.19 |
| 4 | 2087 | 0.0511 | 304 | 0.15 | 403 | 0.19 |
| 3 | 2617 | 0.0641 | 404 | 0.15 | 522 | 0.20 |
| 2 | 3442 | 0.0844 | 577 | 0.18 | 712 | 0.21 |
| 1 | 5050 | 0.1238 | 968 | 0.19 | 1116 | 0.22 |

The capacity of our RHC algorithm is affected by $RR_{Th}$, which is the threshold to determine the significant coefficients. The smaller $RR_{Th}$ is, the higher the capacity. Table 9.1 shows the effect of $RR_{Th}$ under JPEG compression (QF = 20) and Gaussian noise (SD = 500; mean = 0). We observe that our algorithm can not only enlarge the capacity (i.e., >0.12 BPP) but also maintain a low MER (i.e., <0.22) even under voluminous distortions.

## 9.2 HIGH-CAPACITY MULTIPLE-REGIONS-OF-INTEREST WATERMARKING FOR MEDICAL IMAGES

Medical imaging devices have been widely used to produce numerous pictures of the human body in digital form. The field of telemedicine has grown rapidly in recent decades with the rise of telecommunication and information technologies, which has eliminated distance barriers and provided immediate access to medical services all over the world. Telemedicine requires improved digital content protection as multimedia data is widely distributed across the Internet [18–20]. Copyright protection, broadcast monitoring, and data authentication are important issues of information security when medical images are exchanged between hospitals and medical organizations. Hence, image watermarking [21–25] is frequently applied as a process to embed data, such as signature images or textual information, into an image.

In general, two domains are used in image watermarking: the spatial and frequency domains. In the spatial domain, many methods of watermarking have been proposed, such as the difference expansion approach, which can embed bits by utilizing correlations between pixels [26]. To increase the embedding capacity and decrease the distortion of the watermarked image for human perception, the difference expansion method was developed using a generalized integer transform [27] and embedding in the least-significant bits [28,29]. However, since data are inserted into the spatial domain of image pixels, the embedded information is relatively easy to be detected by computer analysis. In frequency domain methods [30–32], a watermark is embedded into the coefficients of the frequency domain of an image. The most frequently used transforms include the DFT, DWT, and DCT. Consequently, the embedded data are distributed around the entire image, which makes the watermark difficult to detect. However, the capacity of the frequency domain methods is often lower than that of the spatial domain methods.

Since the quality of medical images is critical for medical diagnosis, we need to preserve image quality while increasing the embedding capacity. Having this purpose in mind, we consider region of interest (ROI)-based embedding methods [33–38]. An ROI contains critical information for diagnosis and must be saved without any distortion. The remaining *region of noninterest* (RONI) can be utilized for watermark embedding, which allows lossy compression. Our previous work [36] was focused on single-ROI watermarking and the correction of rounding errors, which are deviations that occur when converting real numbers to integers. In this chapter, we focus on multiple RONI medical image watermarking and propose a novel scheme to achieve high data capacity and high image quality in the frequency domain.

This section is organized as follows: the proposed technique is presented in Section 9.2.1, and experimental results and analysis are described in Section 9.2.2.

### 9.2.1 PROPOSED TECHNIQUE

The ROI is critical in providing medical diagnostic information for health-care providers. To secure medical images through watermarking, the ROI must be preserved, so that watermark embedding is only applied to the remaining part of the image, the RONI. In order to enhance the embedding capacity without distorting the important diagnostic information, we take medical images with multiple ROIs preselected by

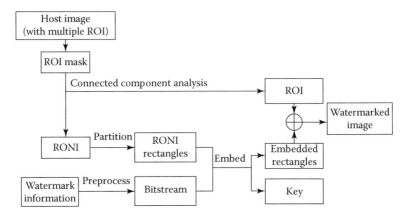

**FIGURE 9.16**   The overall embedding process.

experts as the input and keep ROIs lossless. The remaining RONIs are used for watermark embedding and allow lossy compression. The ROIs are restricted to rectangular shapes due to the image transformation being applied. For an arbitrarily shaped ROI, the smallest bounding box is used.

The RONI part constitutes a concave rectangle. First, the ROI and RONI are obtained through connected-component analysis [25]. Second, the concave RONI rectangle is decomposed into a minimum number of non-overlapping rectangles (i.e., partitioning), which are then transformed into the frequency domain. Third, we perform our embedding algorithm on the preprocessed watermark bitstream along with each RONI rectangle to generate an embedded rectangle and a corresponding key. Note that different types of watermarks can be embedded for robustness. Finally, the watermarked image is obtained by combining these embedded rectangles and the ROI. Figure 9.16 shows the overall embedding process.

During the decoding process, we obtain the ROI and RONI from the watermarked image, collate the RONI rectangles, and perform our decoding algorithm on each RONI rectangle along with a key to extract the original rectangle and the bitstream. We obtain the original image by combining the generated rectangles and the ROI, and retrieve the original watermark by reconstructing the bitstream. Figure 9.17 shows the overall decoding process.

### 9.2.1.1   Watermark Information Preprocessing and Reconstruction

The purpose of preprocessing watermark information is to generate a bitstream and enhance robustness. For medical images, we consider watermark images and textual data.

#### 9.2.1.1.1   Signature Image

For watermark images, *set partitioning in hierarchical trees* (SPIHT) compression [24] is adopted for preprocessing. SPIHT is an image compression algorithm using an essential bit allocation strategy that produces a progressively embedded scalable

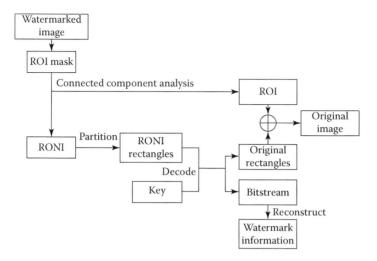

**FIGURE 9.17** The overall decoding process.

00000010100110000000000010000000000000000000011011110000000000000000000000000001110001
00000000000011000010110000101100101100100100101000000010110010000000100011010101000
00010110000000000001011001010000000101100001000001101010010001000110010001000100
10100000101010101010101001011010100100000001000110010101010001100001010000000010
00010100000000010000000000000000001001110010101110100011100010101001101001110010
11001010001000111000100010011111100011010010010100011101010101011001100100000110
1001100111101010011001101010100100001101001100100011111101111111100100001111110010
11010110000101100010010010000001000001011000000000010010100100010000110
(a)

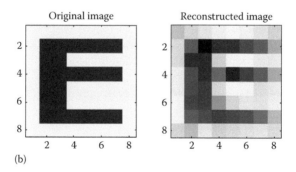

**FIGURE 9.18** An example of SPIHT: (a) the bitstream by SPIHT, (b) original and reconstructed images.

bitstream. The algorithm first encodes the most essential DWT coefficients, followed by transmitting the bits. Therefore, an increasingly refined compression of the original image can be obtained progressively. An example of SPIHT is shown in Figure 9.18, where (a) lists the bitstream generated by applying SPIHT and (b) shows the original and reconstructed images.

### 9.2.1.1.2   Textual Data

Many techniques have been developed for encrypting textual data. The main idea is to translate plain text into secret codes of cipher text. An easy way to achieve encryption is bit shifting. That is, we consider a character to be a byte and shift its position. In this chapter, we adopt an encryption method by taking a logarithm of ASCII codes [20] for the purpose of robustness. The encryption algorithm can be mathematically stated as

$$T_e = \log(T_o \times 2) \times 100 - 300, \qquad (9.5)$$

where $T_e$ denotes the encrypted text and $T_o$ denotes the ASCII code of the original text. The decrypted text can be obtained by

$$T_o = \exp\{\frac{T_e + 300}{100} - \log 2\}. \qquad (9.6)$$

Note that $T_e$ is stored as an integer.

### 9.2.1.2   RONI Partitioning Algorithm

Given a host image with multiple preselected rectangular ROIs, a binary ROI mask can be simply generated by placing 1s inside each ROI rectangle and 0s inside the RONI part. Then, connected-component analysis is applied to locate the position of each ROI rectangle by finding each area connected to a 1. In Figure 9.19, (a) shows an image with two preselected ROIs and (b) shows the generated ROI mask. The upper-left corners of the rectangles are (24,102) with a width of 96 and a height of 129 for the left ROI, and (142,45) with a width of 95 and height of 128 for the right ROI. The location of each ROI is recorded.

The concave RONI part shown in black in Figure 9.19b is decomposed into a number of non-overlapping parts. We partition the concave RONI part into a minimum number of non-overlapping rectangles and transform them from the spatial to the frequency domain. It has been proved by Chadha and Allison [39] that for a rectilinear polygon $M$, the minimum number of rectangles $P$ that $M$ can be partitioned into is

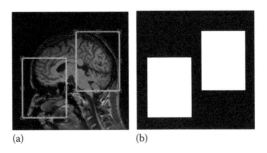

(a)                          (b)

**FIGURE 9.19**   An example of ROI mask generation.

$$P = \frac{n}{2} + h - 1 - C, \tag{9.7}$$

where:

$n$    denotes the total number of vertices of the holes and $M$ itself

$h$    denotes the number of holes

$C$   denotes the maximum cardinality of a set $S$ of concave lines, no two of which are intersected

Furthermore, we develop a novel algorithm for partitioning a RONI concave rectangle into a minimum number of rectangles with the coordinates of all vertices. This algorithm first draws horizontal and vertical concave lines respectively for each ROI, and then selects the result of minimum rectangles as the partition.

Let $[(x_1,y_1),(x_2,y_2),(x_3,y_3),(x_4,y_4)]$ respectively denote the coordinates of the upper-left, upper-right, lower-left, and lower-right corners of each ROI. The algorithm can be stated as Algorithm 1.

*Algorithm 1: RONI concave rectangle partitioning*

Input: $[(x_1,y_1),(x_2,y_2),(x_3,y_3),(x_4,y_4)]$ of each ROI

Output: Minimum number of rectangles partition

1. While $x_1$, $x_3 \neq$ image margin,

$$x_{1\_end} = x_1 - 1; \; x_{3\_end} = x_3 - 1.$$

If $(x_{1\_end}, x_{3\_end}$ go through other ROIs), break.

2. While $x_2, x_4 \neq$ image margin,

$$x_{2\_end} = x_2 + 1; \; x_{4\_end} = x_4 + 1.$$

If $(x_{2\_end}, x_{4\_end}$ go through other ROIs), break.

3. Draw lines from point $(x_{1\_end}, y_1)$ to point $(x_{2\_end}, y_2)$ and from point $(x_{3\_end}, y_3)$ to point $(x_{4\_end}, y_4)$ to generate the horizontal partitioned results.

4. While $y_1, y_2 \neq$ image margin,

$$y_{1\_end} = y_1 - 1; \; y_{2\_end} = y_2 - 1.$$

If $(y_{1\_end}, y_{2\_end}$ go through other ROIs), break.

5. While $y_3, y_4 \neq$ image margin,

$$y_{3\_end} = y_3 + 1; \; y_{4\_end} = y_4 + 1.$$

If $(y_{2\_end}, y_{4\_end}$ go through other ROIs), break.

6. Draw lines from point $(x_1, y_{1\_end})$ to point $(x_3, y_{3\_end})$ and from point $(x_2, y_{2\_end})$ to point $(x_4, y_{4\_end})$ to generate the vertical partitioned results.
7. Select the result from steps 3 and 6 that generates fewer rectangles as the final partition result.

Figure 9.20 shows an example of partitioning the RONI concave rectangle of Figure 9.19b into a minimum number of rectangles. In Figure 9.20, (a) shows the possible horizontal partition, (b) shows the possible vertical partition, and (c) shows the corresponding final result. There are no collinear lines in this example and the horizontal partition is used as a default. In addition, each RONI rectangle is represented as having 1 in the binary partitioned mask. Note that the white areas in Figure 9.20c are the RONI partitioned rectangles for embedding, whereas the white areas in Figures 9.20a,b are the ROIs for preservation. According to Equation 9.7, we obtain $n = 12$, $h = 2$, and $C = 0$. Therefore, $P = 7$ is the optimal partition. Figure 9.6 shows another example of partitioning a three-ROI image with two collinear lines. In this case, we obtain $n = 16$, $h = 3$, and $C = 2$, and therefore $P = 8$ Figure 9.21.

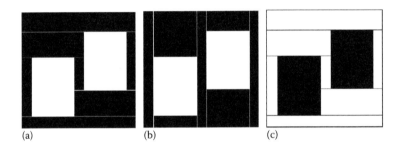

(a)                    (b)                    (c)

**FIGURE 9.20**  An example of partitioning.

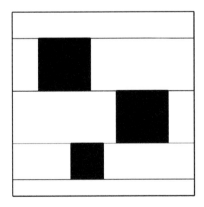

**FIGURE 9.21**  An example of partitioning a three-ROI image with two collinear lines.

### 9.2.1.3 Algorithm for Embedding and Decoding

After the RONI partitioned rectangles and the secret bitstream data are obtained, we can embed the bitstream into each rectangle. For each partitioned rectangle, we transform it into the frequency domain using the DCT. Since the DCT is more computationally effective and tends to have more energy concentrated in fewer coefficients when compared with other transforms such as the DFT, it can break up frequency bands for watermark embedding. Shapiro [40] specifically compared DFT, DWT, and DCT performance. The DCT, the cosine part of the DFT, is asymmetric and has less energy in higher-frequency coefficients than the DFT. Thus, it allows watermark embedding without creating visible boundary artifacts. The DFT is often used for general spectral analysis to find ways into a range of fields. The DWT is the time–frequency resolution of an image. In this chapter, as the clear frequency bands feature is preferable, we selected the DCT.

The *middle frequency area* (MFA) is chosen for embedding because it can avoid the most important parts for human perception (the low-frequency components) and protect the watermark from removal by compression and noise (high-frequency components). Since images follow the power law [41,42], the low-frequency areas convey most of the energy in an image. Human vision is often focused on those most energetic parts. Specifically, in medical images, which are viewed as piecewise smooth or piecewise constant, the magnitude spectrums share almost the same curve because the energy is concentrated in lower frequencies. Figure 9.22 shows an example of the magnitude spectrum of a medical image in a log scale.

In the literature, the MFA is obtained by setting thresholds, and these thresholds can vary with the application. Through the DCT, low-frequency components are

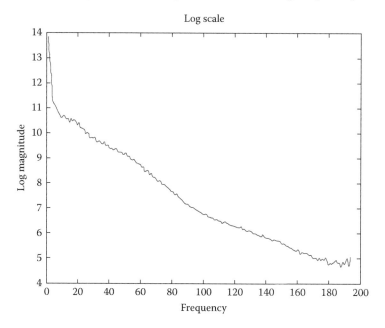

**FIGURE 9.22** The magnitude spectrum.

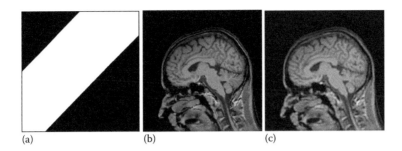

(a)                              (b)                          (c)

**FIGURE 9.23**  The MFA of an image.

placed in the upper-left corner, and high-frequency components are in the lower-right corner. We set the low-frequency area as a triangle with two sides being half of the image's width and height, and the high-frequency area as a triangle with two sides being four-fifths of the image's width and height. The remaining part is the MFA, shown as the white area of Figure 9.23a. Figure 9.23b shows a medical image, and Figure 9.23c is the reconstructed image after MFA removal. It is observed that the MFA removal does not cause much distortion to human vision.

After selecting the MFA, we perform the embedding algorithm for the bitstream. We exploit the amplitude of the original image's energy, indicated by the coefficients of the frequency domain for embedding. After embedding, the amplitude in the MFA is slightly weakened. The number of embedding loops $t$ determines the weakening level. The detailed embedding algorithm is presented as Algorithm 2. Note that if rounding errors occur in the DCT process, we can correct them by applying our previous work in [36].

*Algorithm 2: Embedding algorithm*

Input: MFA, bitstream $S$, and the number of embedding loops $t$

Output: Embedded MFA and a key $K$

1. Obtain the mean value $M$ of the MFA.
2. For each coefficient $V$ in the MFA, compute middle point $A$ by calculating the arithmetic mean of $V$ and $M$. Store $A$ as the key $K$. Note that we can perform the encryption algorithm described in Section 9.2.1.1 on $K$.
3. Compute the embedding length $L$ by

$$L = |\,\text{round}(\log|A - M|)\,|. \tag{9.8}$$

Note that if $L = 0$ or $|A - M| <$ the logarithm base, 1 bit will be embedded.

4. Obtain the embedding integer value $E$ by taking $L$ bits of $S$.
5. If ($V$ is positive), update $V = V + E$,

else update $V = V - E$.

6. Repeat steps 1 through 5 $t$ times.

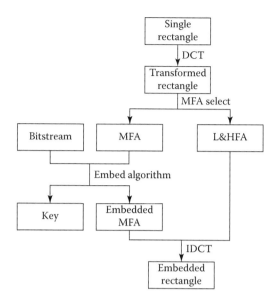

**FIGURE 9.24**  The detailed embedding algorithm.

After the embedding algorithm is applied, we generate the embedded MFA. The whole rectangle can be obtained by combining the embedded MFA with the low- and high-frequency areas (L&HFA). Then, the embedded rectangle is obtained by applying the IDCT to the whole rectangle. Figure 9.24 shows the process of embedding. Finally, all the embedded RONI rectangles are combined with the ROIs to form the stego-image (i.e., the watermarked image).

For watermark extraction, we transform a RONI partitioned rectangle into the frequency domain, select the MFA, perform the reverse decoding algorithm, and combine all the decoded rectangles to obtain the original image. Figure 9.25 shows the detailed decoding process. The decoding algorithm is presented as Algorithm 3. Note that we can retrieve the watermark with the reconstruction algorithm described in Section 9.2.1.1.

*Algorithm 3: Extraction algorithm*
Input: MFA′ of a stego-image, key $K$, and the number of embedded loops $t$.
Output: Original MFA and the watermark bitstream $B$.

1. Obtain the mean value $M'$ of the MFA″.
2. For each coefficient $V'$ in the MFA″, compute the embedded value $E'$ by

$$E' = |V' - K|. \tag{9.9}$$

3. Compute the embedded length $L'$ by

$$L' = |\,\mathrm{round}(\log|K - M'|)\,|. \tag{9.10}$$

Note that if $L' = 0$ or $|K - M'| <$ the logarithm base, 1 bit will be extracted.

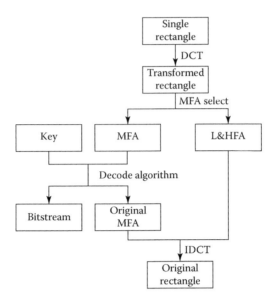

**FIGURE 9.25**    The detailed decoding algorithm.

  4. Obtain the embedded bits stream $B'$ by converting $E'$ with length $L'$.

  5. For each coefficient $V'$ in the MFA″, update $V'$ by

$$V' = 2K - M'. \tag{9.11}$$

  6. Obtain the original MFA by repeating steps 1 to 5 $t$ times and obtain the whole bitstream $B$ by stacking $B'$.

### 9.2.2 EXPERIMENTAL RESULTS

Figure 9.26 shows an example of embedding a watermark image into an MRI brain image with multiple ROIs and setting the number of embedding loops $t$ to 1. In Figure 9.26, (a) is the medical image of size $427 \times 600$ with six ROIs, (b) shows the RONI partition with 17 rectangles, (c) is the watermarked image, (d) is the difference between the original and watermarked images, and (e) shows the original and reconstructed watermarks. Note that the differences are multiplied by 100 to enhance the highlights. Also note that the six ROIs are lossless.

The biggest reason for the differences appearing at the image's edge is that we use the RONIs, as shown by the white area of Figure 9.26b, for watermark embedding. In other words, the differences depend on the locations of the RONIs. The preselected diagnosis parts are in the middle, so the RONIs are at the edges of the host image. Furthermore, since we have a high capacity, only the top few RONI rectangles are sufficient to embed the watermark. The quality of the reconstructed image is lower because the original watermark is compressed. The embedded content is actually a compressed or encrypted bitstream. We reconstruct the watermark image with the extracted bitstream.

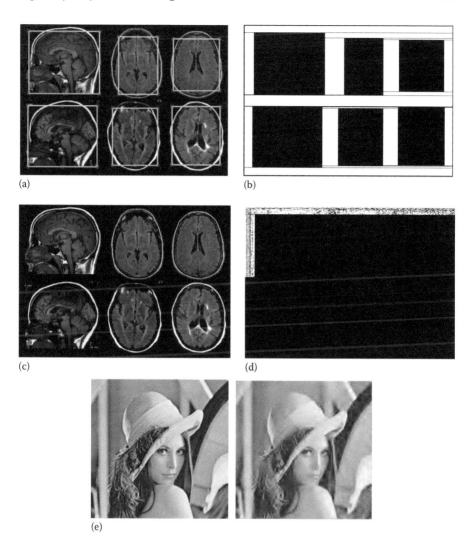

**FIGURE 9.26** An example of embedding a watermark image: (a) a medical image with six ROIs, (b) RONI rectangle partition, (c) the watermarked image, (d) the difference between the original and watermarked images (amplified 100 times), (e) the original and reconstructed watermarks.

The textual information shown in Figure 9.27 is also tested. By setting the number of embedding loops $t$ to 1, both the image and text embedding results are listed in Table 9.2. The encrypting bitstream length is the number of bits for embedding. The embed area is the number of pixels for embedding. The *peak signal-to-noise ratio* (PSNR) is defined as

$$PSNR(f, g) = 10 \times \log_{10} \frac{(2^k - 1)^2}{MSE}, \tag{9.12}$$

```
The patient information

Patient ref. no: AX8865098

Name of the doctor: Dr.X

Name of the doctor: Dr.C

Age: 48

Address: 22 midland ave

Case history:

Date of admission: 09.20.2015

Results: T wave inversion

Diagnosis: Suspected MI
```

**FIGURE 9.27**   Embedded textual information.

**TABLE 9.2**
**Bits Length, Embed Area, Capacity, and PSNR**

| Watermark Type | Bitstream Length (bits) | Embed Area (pixels) | Total Capacity(bits) | PSNR(dB) |
|---|---|---|---|---|
| Image | 11920 | 54092 | 86692 | 48.72 |
| Text | 2552 | 54092 | 86692 | 62.48 |

where $k$ is the bit depth and *MSE* is the mean square error.

$$MSE = \frac{1}{mn} \sum_{1}^{m} \sum_{1}^{n} \| f(i,j) - g(i,j) \|^2. \tag{9.13}$$

Table 9.2 implies that for $t = 1$, the BPP of the proposed technique is about 1.6. We can increase $t$ to achieve a higher BPP; however, the PSNR would decrease slightly.

To further explore the relationship between BPP and PSNR in medical image watermarking, we compute the BPP and PSNR in a $256 \times 256$ MRI brain image, as shown in Figure 9.28, where the embed area is 36,293 pixels and $t$ ranges from 1 to 5. Table 9.3 lists the results, indicating that when $t$ increases, BPP increases but PSNR decreases.

We can continue increasing $t$ for a higher BPP, while the PSNR will be compromised. It is observed that when the BPP increases linearly, the PSNR declines exponentially. We also compare our proposed technique with six similar techniques that are currently available. The results are listed in Tables 9.4 and 9.5. Among them, [25], [26], and [29] are the ROI medical image watermarking methods, and [27], [42],

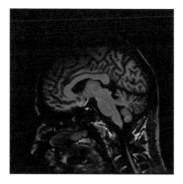

**FIGURE 9.28**    A MRI brain image with an embed area of 36,293 pixels.

**TABLE 9.3**
**PSNR, BPP, and Capacity (Number of Embedding Loops *t* from 1 to 5)**

| *t* | 1 | 2 | 3 | 4 | 5 |
|---|---|---|---|---|---|
| PSNR(dB) | 48.53 | 45.00 | 43.66 | 43.07 | 42.78 |
| BPP | 1.63 | 3.06 | 4.40 | 5.71 | 7.03 |
| Capacity (bits) | 59,138 | 110,956 | 159,552 | 207,336 | 255,052 |

**TABLE 9.4**
**PSNR Comparisons of the Proposed Technique with Existing Techniques**

| Method | PSNR(dB) | Increments Compared with Our PSNR in the *t* Range 48.53 dB to 42.78 dB | | | | |
|---|---|---|---|---|---|---|
| | | *t* | | | | |
| | | 1 | 2 | 3 | 4 | 5 |
| [27] | 29.23 | 66.03% | 53.95% | 49.37% | 47.35% | 46.36% |
| [25] | 38.00 | 27.71% | 18.42% | 14.89% | 13.34% | 12.58% |
| [42] | 29.39 | 65.12% | 53.11% | 48.55% | 46.55% | 45.56% |
| [43] | 31.48 | 54.16% | 42.95% | 38.69% | 36.82% | 35.90% |
| [26] | 22.36 | 117.04% | 101.25% | 95.26% | 92.62% | 91.32% |
| [37] | 51.24 | −5.28% | −12.17% | −14.79% | −15.94% | −16.51% |
| [29] | 31.70 | 53.09% | 41.96% | 37.73% | 35.87% | 34.95% |
| [38] | 44.64 | 8.71% | 0.81% | −2.19% | −3.51% | −4.17% |

*Note:* A negative increment means that our PSNR is lower.

**TABLE 9.5**

**BPP Comparisons of the Proposed Technique with Existing Techniques**

| | | Increments Compared with Our BPP in the $t$ Range 1.63 to 7.03 | | | | |
|---|---|---|---|---|---|---|
| | | | | $t$ | | |
| Method | BPP | 1 | 2 | 3 | 4 | 5 |
| [27] | 0.74 | 120.27% | 313.51% | 494.59% | 671.62% | 850.00% |
| [25] | 1.00 | 63.00% | 206.00% | 340.00% | 471.00% | 603.00% |
| [42] | 0.99 | 65.65% | 209.09% | 344.44% | 476.76% | 610.10% |
| [43] | 0.49 | 232.65% | 524.49% | 797.96% | 1065.31% | 1334.69% |
| [26] | 2.00 | -18.50% | 53.00% | 120.00% | 185.5% | 251.5% |
| [37] | 0.54 | 201.85% | 466.67% | 714.81% | 957.41% | 1201.85% |
| [29] | 1.06 | 53.77% | 188.68% | 315.09% | 438.68% | 563.21% |
| [38] | 0.22 | 640.91% | 1290.90% | 1900.00% | 2495.45% | 3095.45% |

*Note:* A negative increment means that our BPP is lower.

[43], [37], and [38] are the medical image-watermarking methods without ROI but with similar embedding strategies for various usages.

By limiting $t$ from 1 to 5, as compared with [27], the proposed technique can improve the PSNR from 46.36% to 66.03% and BPP from 120.27% to 850%. As compared with [36], the proposed technique can improve the PSNR from 12.58% to 27.71% and BPP from 63% to 603%. As compared with [43], the proposed technique can improve the PSNR from 45.56% to 65.12% and BPP from 65.65% to 610.10%. As compared with [26], the proposed technique can improve the PSNR from 35.90% to 54.16% and BPP from 232.65% to 1334.69%.

As compared with [37], the proposed technique can improve the PSNR from 91.32% to 117.04%. Although the BPP is 18.50% lower than what it is for $t = 1$, it could be increased by 251.5% if $t$ is increased to 5. As compared with [44], the proposed technique can improve the PSNR from $-16.51$% to $-5.28$% and BPP from 201.85% to 1201.85%. Although our PSNR is at most 16.51% lower, the capacity is increased up to 1201.85%.

As compared with [29], the proposed technique can improve the PSNR from 34.95% to 53.07% and BPP from 53.77% to 563.21%. Compared with [38], the proposed technique can improve the PSNR from $-4.17$% to 8.71% and BPP from 640.91% to 3095.45%. When $t > 3$, we have a lower PSNR, but the capacity is increased by 3095%. These comparisons indicate that the proposed technique has achieved a significantly high quality and high capacity. In addition, the comparison is conducted when a watermark is embedded into a $256 \times 256$ MRI brain image, which produces a higher PSNR when preselected ROIs are used.

In summary, the proposed technique keeps ROIs lossless and enhances both the watermarking capacity and quality in RONIs. Thus, it maintains quality while simultaneously increasing the capacity for medical image watermarking. Furthermore, our technique allows flexible adjustment to the variable $t$ that controls the trade-off between image fidelity and embedding capacity.

## REFERENCES

1. Berghel, H. and O'Gorman, L., Protecting ownership rights through digital watermarking, *IEEE Computer Mag.*, 101, 1996.
2. Eggers, J. and Girod, B., *Informed Watermarking*, Kluwer Academic, Norwell, MA, 2002.
3. Wu, Y. T., Multimedia security, morphological processing, and applications, PhD dissertation, New Jersey Institute of Technology, Newark, 2005.
4. Nikolaidis, N. and Pitas, I., Robust image watermarking in the spatial domain, *Signal Processing*, 66, 385, 1999.
5. Celik, M. U. et al., Hierarchical watermarking for secure image authentication with localization, *IEEE Trans. Image Processing*, 11, 585, 2002.
6. Voyatzis, G. and Pitas, I., Applications of toral automorphisms in image watermarking, in *Proc. IEEE Int. Conf. Image Processing*, Lausanne, Switzerland, 1996, 237.
7. Mukherjee, D. P., Maitra, S., and Acton, S. T., Spatial domain digital watermarking of multimedia objects for buyer authentication, *IEEE Trans. Multimedia*, 6, 1, 2004.
8. Wong, P. W., A public key watermark for image verification and authentication, in *Proc. IEEE Int. Conf. Image Processing*, Chicago, IL, 1998, 425.
9. Cox, I. J. et al., Secure spread spectrum watermarking for multimedia, *IEEE Trans. Image Processing*, 6, 1673, 1997.
10. Zhao, D., Chen, G., and Liu, W., A chaos-based robust wavelet-domain watermarking algorithm, *Chaos, Solitons and Fractals*, 22, 47, 2004.
11. Miller, M. L., Doerr, G. J., and Cox, I. J., Applying informed coding and embedding to design a robust high-capacity watermark, *IEEE Trans. Image Processing*, 13, 792, 2004.
12. Wu, Y. T. and Shih, F. Y., An adjusted-purpose digital watermarking technique, *Pattern Recognition*, 37, 2349, 2004.
13. Shih, F. Y. and Wu, Y., Enhancement of image watermark retrieval based on genetic algorithm, *J. Visual Communication and Image Representation*, 16, 115, 2005.
14. Lin, S. D. and Chen, C.-F., A robust DCT-based watermarking for copyright protection, *IEEE Trans. Consumer Electronics*, 46, 415, 2000.
15. Wu, Y. T. and Shih, F. Y., Genetic algorithm based methodology for breaking the steganalytic systems, *IEEE Trans. Systems, Man, and Cybernetics: Part B*, 36, 24, 2006.
16. Shih, F. Y. and Wu, Y. T., Combinational image watermarking in the spatial and frequency domains, *Pattern Recognition*, 36, 969, 2003.
17. Wang, H., Chen, H., and Ke, D., Watermark hiding technique based on chaotic map, in *Proc. IEEE Int. Conf. Neural Networks and Signal Processing*, Nanjing, China, 2003, 1505.
18. Chen, B., Shu, H., Coatrieux, G., Chen, G., Sun, X., and Coatrieux, J. L., Color image analysis by quaternion-type moments, *J. Mathematical Imaging and Vision*, 51, 124, 2015.
19. Huang, J., Shi, Y., and Shi, Y., Embedding image watermarks in DC components, *IEEE Trans. Circuits and Systems for Video Technology*, 10, 974, 2000.
20. Ruderman, D. L. and Bialek, W., Statistics of natural images: Scaling in the woods, *Physical Review Letters*, 73, 814, 1994.
21. Berghel, H. and O'Gorman, L., Protecting ownership rights through digital watermarking, *Computer*, 29, 101, 1996.
22. Dittmann, J. and Nack, F., Copyright-copywrong, *IEEE Multimedia*, 7, 14, 2000.
23. Eswaraiah, R. and Sreenivasa Reddy, E., Medical image watermarking technique for accurate tamper detection in ROI and exact recovery of ROI, *Int. J. Telemedicine and Applications*, 2014.

24. Guo, P., Wang, J., Li, B., and Lee, S., A variable threshold-value authentication architecture for wireless mesh networks, *J. Internet Technology*, 15, 929, 2014.
25. Shih, F. Y. and Wu, Y. T., Robust watermarking and compression for medical images based on genetic algorithms, *Information Sciences*, 175, 200, 2005.
26. Wakatani, A., Digital watermarking for ROI medical images by using compressed signature image, in *Proc. 35th IEEE Annual Hawaii Int. Conf. System Sciences*, 2043, 2002.
27. Alattar, A. M., Reversible watermark using the difference expansion of a generalized integer transform, *IEEE Trans. Image Processing*, 13, 1147, 2004.
28. Al-Qershi, O. M. and Khoo, B.E., Two-dimensional difference expansion (2D-DE) scheme with a characteristics-based threshold, *Signal Processing*, 93, 154, 2013.
29. Zain, J. M. and Clarke, M., Reversible region of non-interest (RONI) watermarking for authentication of DICOM images, *Int. J. Computer Science and Network Security*, 7, 19, 2007.
30. Bas, P., Chassery, J. M., and Macq, B., Image watermarking: An evolution to content based approaches, *Pattern Recognition*, 35, 545, 2002.
31. Lin, S. D. and Chen, C.F., A robust DCT-based watermarking for copyright protection, *IEEE Trans. Consumer Electronics*, 46, 415, 2000.
32. Loganathan, R. and Kumaraswamy, Y. S., Medical image compression with lossless region of interest using adaptive active contour, *J. Computer Science*, 8, 747, 2012.
33. Benyoussef, M., Mabtoul, S., Marraki, M., and Aboutajdine, D., Robust ROI watermarking scheme based on visual cryptography: Application on mammograms, *J. Information Processing Systems*, 11, 495, 2015.
34. Gao, X., An, L., Yuan, Y., Tao, D., and Li, X., Lossless data embedding using generalized statistical quantity histogram, *IEEE Trans. Circuits and Systems for Video Technology*, 21, 1061, 2011.
35. Rajendra Acharya, U., Acharya, D., Subbanna Bhat, P., and Niranjan, U. C. Compact storage of medical images with patient information, *IEEE Trans. Information Technology in Biomedicine*, 5, 320, 2001.
36. Srivastava, A., Lee, A. B., Simoncelli, E. P., and Zhu, S. C., On advances in statistical modeling of natural images, *J. Mathematical Imaging and Vision*, 18, 17, 2003.
37. Wang, Z. H., Lee, C. F., and Chang, C. Y., Histogram-shifting-imitated reversible data hiding, *J. Systems and Software*, 86, 315, 2013.
38. Zhao, Z., Luo, H., Lu, Z. M., and Pan, J. S., Reversible data hiding based on multilevel histogram modification and sequential recovery, *AEU Int. J. Electronics and Communications*, 65, 814, 2011.
39. Chadha, R. and Allison, D., Decomposing rectilinear figures into rectangles, Technical Report, Department of Computer Science, Virginia Polytechnic Institute and State University, Blacksburg, 1988.
40. Shapiro, J. M., Embedded image coding using zerotrees of wavelet coefficients, *IEEE Trans. Signal Processing*, 41, 3445, 1993.
41. Shaamala, A., Abdullah, S. M., and Manaf, A. A., Study of the effect DCT and DWT domains on the imperceptibility and robustness of genetic watermarking, *Int. J. Computer Science Issues*, 8, 2011.
42. Thodi, D. M. and Rodríguez. J. J., Expansion embedding techniques for reversible watermarking, *IEEE Trans. Image Processing*, 16, 721, 2007.
43. Tian, J., Reversible data embedding using a difference expansion, *IEEE Trans. Circuits and Systems for Video Technology*, 13, 890, 2003.
44. Xia, Z., Wang, X., Sun, X., Liu, Q., and Xiong, N., Steganalysis of LSB matching using differences between nonadjacent pixels, *Multimedia Tools and Applications*, 2014.

# 10 Reversible Watermarking

## 10.1 REVERSIBLE IMAGE AUTHENTICATION SCHEME BASED ON CHAOTIC FRAGILE WATERMARK

With the recent rapid development of information technology, digital multimedia is being widely used in many applications. However, the rise of information security problems, such as malicious alteration and forgery, has become increasingly serious. Fragile digital watermarking is used to accomplish data authentication, and a variety of fragile watermark-based image authentication schemes have been developed. Among them, the seminal fragile watermarking scheme proposed by Yeung and Mintzer [1] is based on the parity of pixel values and uses error diffusion to maintain the average color. Fridrich [2] and Holliman and Memon [3] analyzed its security drawbacks and proposed attacks and improvements, respectively. Since the function of the fragile watermarking scheme is to authenticate the integrity of the media, the watermark should reflect the content of the media, which is quite different from its counterpart in the robust watermarking scheme.

The hierarchical block-based watermarking scheme proposed by Celik et al. [4] not only includes information from the corresponding block but also possesses relative information from higher-level blocks. Chan and Chang [5] introduced the *(7,4) Hamming* code into the image authentication arena. Liu et al. [6] employed permutation transformation and chaotic image patterns to break the corresponding positional relation between pixels in the watermarked image and those in the watermark. Zhang and Wang [7] proposed a fragile watermarking scheme with a hierarchical mechanism, in which the embedded watermark data are derived from both pixels and blocks. Chen and Lin [8] presented a method of finding nearly optimal positions for embedding authentication messages using genetic algorithms to improve the quality of protected images. Luo et al. [9] proposed a dynamic compensation method against *least-significant-bit* (LSB) steganalysis. Note that most existing fragile watermarking schemes leave some permanent distortions in the original image for embedding the watermark information. However, for some multimedia applications, such as medical diagnosis, law enforcement, and fine art, the generated distortions are not allowed.

Reversible data-hiding schemes have also been proposed. Tian [10] proposed a difference expansion data-hiding scheme, where the difference and average values of two neighboring pixels are calculated and the secret data to be embedded are appended to a difference value represented as a binary number. Alattar [11], Kim et al. [12], and Weng [13] further extended Tian's work. Ni et al. [14] utilized the histogram of the pixels in the cover image to design a reversible hiding scheme, where the pixels between the peak and zero pair are modified in the embedding process and the pixel in the peak point is used to carry a bit of the secret message. Some authors, such as Lin et al. [16], have proposed methods to increase the hiding capacity. Chen and Kao [16] proposed a reversible image-watermarking approach using quantized DCT coefficients. Another data-hiding scheme based on reversible contrast mapping was proposed by Coltuc and Chassery [17]. It can efficiently achieve a large hiding capacity without any additional data compression. The inherent relationship between data hiding and watermarking motivates us to integrate the reversible data-hiding techniques with the fragile watermark-based image authentication approaches.

In this section, we present a novel fragile watermark-based reversible image authentication scheme. The chaotic hash value of each image block is computed as the watermark, which ensures that there exists complicated nonlinear and sensitive dependence within the image's gray features, secret key, and watermark. The corresponding size of the key space is hugely beneficial for image authentication. Moreover, reversible watermark embedding allows the exact recovery of the original image after the authentication message has been extracted.

The watermarking scheme is introduced in Section 10.1.1. A performance analysis is given in Section 10.1.2. Finally, a conclusion is given in Section 10.1.3.

### 10.1.1 WATERMARKING SCHEME

#### 10.1.1.1 Embedding Algorithm

1. *Partition an image of size $512 \times 512$ into non-overlapping blocks of size $64 \times 64$.*
2. *Compute the chaotic watermark of each block.*

The hash value of each block is generated as the corresponding watermark. If a conventional hash function is used, the message has to be padded in such a manner that its length is a multiple of a specific integer. While considering the inherent merits of chaos, such as one-way sensitivity to tiny changes in initial conditions and parameters, we can fulfill this in a simple way without any extra padding. Here, we use a chaotic method to compute the hash value of each block, which is derived from the chaotic hash functions of text messages [18–22].

We use the *piecewise linear chaotic map* (PWLCM) as defined by

$$X_{t+1} = F_P(X_t) = \begin{cases} X_t / P, & 0 \le X_t < P \\ (X_t - P) / (0.5 - P), & P \le X_t < 0.5 \\ (1 - X_t - P) / (0.5 - P), & 0.5 \le X_t < 1 - P \\ (1 - X_t) / P, & 1 - P \le X_t \le 1 \end{cases} \quad (10.1)$$

where $X_t \in [0,1]$ and $P \in (0,0.5)$ denote the iteration trajectory value and the current iteration parameter of the PWLCM, respectively. The gray-level range in [0,255] is linearly mapped into [0,1] and then modulated into the chaotic iteration parameter and trajectory. We describe more details as follows.

a. The image block of size $64 \times 64$ is converted into a 1D array $C$ in a raster scan of left to right and top to bottom, and their gray values are normalized, so each element $C_i \in [0,1]$ and the length $s = 64 \times 64$.

b. Perform the following iteration process of the PWLCM. 1st iteration:

$$P_1 = (C_1 + P_0)/4 \in (0,0.5), \ X_1 = F_{P_1}(X_0) \in (0,1).$$

2nd~$s$th iterations:

$$P_k = (C_k + X_{k-1})/4 \in (0,0.5), \ X_k = F_{P_k}(X_{k-1}) \in (0,1).$$

$(s+1)$th iteration:

$$P_{s+1} = (C_s + X_s)/4 \in (0,0.5), \ X_{s+1} = F_{P_{s+1}}(X_s) \in (0,1).$$

$(s+2)$th~$2s$th iterations:

$$P_k = (C_{(2s-k+1)} + X_{k-1})/4 \in (0,0.5), \ X_k = F_{P_k}(X_{k-1}) \in (0,1).$$

$(2s+1)$th~$(2s+2)$th iterations:

$$X_k = F_{P_{2s}}(X_{k-1}) \in [0,1].$$

The initial $X_0 \in [0,1]$ and $P_0 \in (0,1)$ of the PWLCM are used as the algorithm keys. Note that the PWLCM in Equation 10.1 is iterated to obtain the next value $X_k$ under the control of the current initial value $X_{k-1}$ and the current parameter $P_k$. If $X_k$ is equal to 0 or 1, then an extra iteration needs to be carried out.

c. Transform $X_{2s}$, $X_{2s+1}$, $X_{2s+2}$ to the corresponding binary format. We extract 40, 40, and 48 bits after the decimal point, respectively. Note that the three numbers are multiples of 8 in a nondecreasing order. Finally, we juxtapose them from left to right to form the 128-bit chaotic hash value as the watermark of the current block.

3. *Embed the watermark of each block into its top-left sub-block of size $32 \times 32$ in a reversible way.*

Motivated by Coltuc and Chassery [17], we adopt the *reversible contrast mapping* (RCM)-based method to embed the watermark of each $64 \times 64$ block into its top-left $32 \times 32$ sub-block. The RCM is a simple integer transform applied to pairs of pixels in an image. Without loss of generality, let $(x,y)$ be a pair of pixels in an 8-bit image. The forward RCM is defined as

$$x' = 2x - y, \ y' = 2y - x. \tag{10.2}$$

The corresponding inverse RCM transform is defined as

$$x = \left\lceil \frac{2}{3}x' + \frac{1}{3}y' \right\rceil, \; y = \left\lceil \frac{1}{3}x' + \frac{2}{3}y' \right\rceil, \tag{10.3}$$

where $\lceil \cdot \rceil$ is the ceil function. To prevent overflow and underflow, the transformed pair is restricted to a subdomain $D \subset [0,255] \times [0,255]$ defined by

$$0 \le 2x - y \le 255, \; 0 \le 2y - x \le 255. \tag{10.4}$$

To avoid decoding ambiguities, some odd pixel pairs must be eliminated—that is, those odd pixel pairs located on the borders of $D$ and subjected to

$$2x - y = 1, \; 2y - x = 1, \; 2x - y = 255, \; 2y - x = 255. \tag{10.5}$$

Let $D_c$ denote the domain of the transform without the ambiguous odd pixel pairs. The watermark-embedding procedure is described as follows:

a.  Partition the pending top-left $32 \times 32$ image sub-block into multiple pairs of pixels on rows. The other sub-blocks of the $64 \times 64$ image block remain unchanged.
b.  Each pair $(x,y)$ is classified into three different types and applied by one of the three corresponding rules. For simplicity, we denote the three cases as A-, B-, and C-types.

   i.  If $(x,y) \notin D_c$, we classify it into C-type pair, set the LSB of $x$ to 0, treat the original LSB of $x$ as an overhead information bit, and add it behind the 128-bit watermark.
   ii. If $(x,y) \notin D_c$ and their LSB pattern is not $(1,1)$, we classify it into an A-type pair, transform $(x,y)$ into $(x',y')$ using Equation 10.2, set the LSB of $x'$ to 1, and embed a bit of the watermark payload and overhead information in the LSB of $y'$.
   iii. If $(x,y) \in D_c$ and their LSB pattern is $(1,1)$, we classify it into a B-type pair, set the LSB of $x$ to 0, and embed a bit of the watermark payload and overhead information in the LSB of $y$.

The information to be embedded includes the 128-bit watermark of the $64 \times 64$ image block (payload) and the bits for C-type pairs (overhead information). According to experimental results, the number of C-type pairs is usually very few, so that the length of the information to be embedded is less than $2^8$ bits, while the available storage for embedding in the $32 \times 32$ sub-block is much larger than $2^8$ bits.

4.  *Combine all the embedded blocks into a watermarked image of size $512 \times 512$.*

### 10.1.1.2  Extraction Algorithm

1.  *Partition a watermarked image of size $512 \times 512$ into non-overlapping blocks of size $64 \times 64$.*

2. *Extract the watermark from the top-left 32 × 32 sub-block of each block and recover the block in a reversible way.*

    The watermark extraction and the full recovery of the original image block are described as follows:

    a. Partition the pending $32 \times 32$ watermarked sub-block into multiple pairs of pixels in rows.

    b. Each pair $(x',y')$ is applied to one of the following three rules.

        i. If the LSB of $x'$ is 1, we recognize it as an A-type pair, extract the LSB of $y'$ and store it into the extracted watermark payload and overhead information, set the LSBs of $x'$ and $y'$ to 0, and recover the original pair $(x,y)$ by inverse transform in Equation 10.3.

        ii. If the LSB of $x'$ is 0 and the pair $(x',y')$ belongs to $D_c$, we recognize it as a B-type pair, extract the LSB of $y'$, store it into the extracted watermark payload and overhead information, and restore the original pair $(x,y)$ as $(x',y')$ with the LSBs set to 1.

        iii. If the LSB of $x'$ is 0 and the pair $(x',y')$ does not belong to $D_c$, we recognize it as a C-type pair and restore the original pair $(x,y)$ by replacing the LSB of $x'$ with the corresponding true value extracted from the watermark overhead information behind the first 128-bit watermark payload.

    c. The first 128 bits of the extracted information are the extracted watermark of this block, while the other sub-blocks of the $64 \times 64$ image block remain unchanged.

3. *Compute the chaotic watermark of each 64 × 64 recovered block and compare it with the extracted one.*

    a. The watermark of the recovered image block is computed similarly to that in Section 10.1.1.1.

    b. The new watermark of the recovered block and the extracted one are compared in order to authenticate whether this block has been modified or not. If they are the same, then no modification was made in this block; otherwise, this block was modified.

4. *Combine all the recovered blocks into a recovered image of size 512 × 512.*

## 10.1.2 PERFORMANCE ANALYSIS

### 10.1.2.1 Perceptual Quality

Deriving from the forward RCM transform in Equation 10.2, we obtain $x' + y' = x + y$ and $x' - y' = 3(x - y)$. It means that the gray-level average is preserved and the difference is tripled. Note that this perceptual change is only transitory. Since the original image can be exactly recovered after the watermark extraction and authentication of our method, no permanent perceptual distortion is left in the image.

We have evaluated the proposed scheme using extensive experiments. For illustration purposes, we show the *Lena* and *Cameraman* images of size $512 \times 512$ in Figure 10.1.

(a)       (b)

**FIGURE 10.1** Two original images: (a) *Lena*, (b) *Cameraman*.

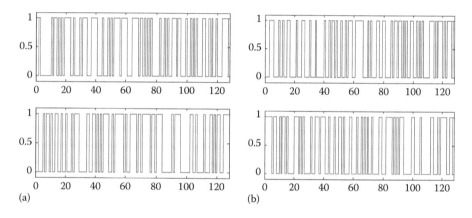

**FIGURE 10.2** 128-bit watermark samples of the first image blocks: (a) *Lena*, (b) *Cameraman*.

Under the control of the algorithm key, all the 128-bit watermarks of the total 64 image blocks have been computed. In experiments, the initial value $X_0$ of the PWLCM is set as 0.232323 and the initial parameter $P_0$ is set as 0.858485. Since there is sensitive dependence between the image's gray value and its watermark, even the slightest modification in gray value, say only LSB modification, will lead to huge changes in the obtained watermark. Figure 10.2 shows the 128-bit watermark samples of the first image blocks with their top-left corner position as (1,1) in *Lena* and *Cameraman*, where the original watermarks are listed on the top and the corresponding watermarks after modifying the LSB of the first pixel are listed on the bottom. The total number of different bits is 65 and 62 for *Lena* and *Cameraman*, respectively. For all the image blocks of size $64 \times 64$, after repeatedly modifying the LSB of the randomly chosen pixel $64 \times 64$ times, we compute the average numbers of the different bits between the original and new watermarks. The average number of different bits for blocks 1, 2, 3, and 4 with their top-left corner position as (1,1), (1,65), (257,193), and (449,449), respectively, are listed in Table 10.1. They are all

**TABLE 10.1**

**Average Numbers of Differences between Original and New Watermarks**

|  | Sample Block 1 | Sample Block 2 | Sample Block 3 | Sample Block 4 |
|---|---|---|---|---|
| *Lena* | 63.8630 | 64.0576 | 64.0520 | 64.0884 |
| *Cameraman* | 64.1140 | 64.0654 | 63.9448 | 63.8801 |

(a)      (b)

**FIGURE 10.3** Two watermarked images: (a) *Lena* with PSNR = 38.0909 dB, (b) *Cameraman* with PSNR = 38.8733 dB.

very close to the ideal 64-bit value—that is, half of the total 128 bits, indicating the sensitive dependence of the watermark on gray value.

By implementing the proposed embedding algorithm, the corresponding watermarked images are shown in Figure 10.3. The *peak signal-to-noise ratios* (PSNRs) between the original and watermarked images are measured by

$$PSNR = 10 \times \log_{10}\left(\frac{255^2}{MSE}\right), \tag{10.6}$$

where MSE is the mean-squared error between the original and watermarked images. For *Lena*, its PSNR is 38.0909 dB, and for *Cameraman*, its PSNR is 38.8733 dB. According to the RCM, there are only some contrast increases.

### 10.1.2.2 Reversible Capability

By implementing the proposed extracting algorithm, the corresponding recovered images are obtained, as shown in Figure 10.4. They are exactly the same as the original counterparts in Figure 10.1. The reversible capability guarantees that no permanent perceptual distortion is left in the image.

### 10.1.2.3 Modification Authentication and Localization Capability

The new watermark of each recovered block and the extracted one are compared to authenticate whether the current block was modified. Since the recovered *Lena*

(a)                              (b)

**FIGURE 10.4**   Two exactly recovered images: (a) *Lena*, (b) *Cameraman*.

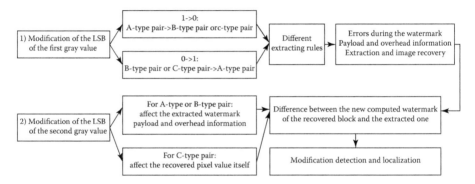

**FIGURE 10.5**   The modification detection and localization procedure.

and *Cameraman* images pass the authentication, it indicates that there have been no modifications to the watermarked images.

In the following, we analyze the modification localization capability of the proposed scheme, as shown in Figure 10.5. Let us take only one LSB modification as an example. Let the LSB pair pattern in the watermarked image be $(x,y)$.

1. If the increment is $(|\Delta x'| = 1, |\Delta y'| = 0)$—that is, the modification happens on the LSB of the first gray value—it means that a former A-type pair turns into a false B-type or C-type pair. Since different type pairs have different extracting rules, the misclassification of the pair type will result in an error during the watermark payload and overhead information extraction as well as during image recovery.
2. If the increment is $(|\Delta x'| = 0, |\Delta y'| = 1)$—that is, the modification happens on the LSB of the second gray value—according to the extracting algorithm in Section 10.1.2, for the A-type or B-type pairs, it will affect the extracted watermark payload and overhead information; for the C-type pair, it will affect the recovered pixel itself.

We have conducted an experiment on the effects of slight modification and the results show that the modification authentication and localization capability of the proposed scheme is guaranteed. For instance, a sample block of size $64 \times 64$ with its top-left corner as (257,193) in the watermarked *Lena* is used to demonstrate the modification localization capability. There are two cases according to the modified location.

1. *Outside the top-left $32 \times 32$ subarea*: A pixel with a gray value of 89 located at (315,216) is randomly chosen. Let the value 89 be changed to 88. Figure 10.6a shows a watermarked *Lena*. By using the proposed algorithm, we can detect and locate the modification, as shown in Figure 10.6b. Actually, this slight modification produces a huge change within the 128-bit watermark of the corresponding image block. Figure 10.7a shows the differences between the new 128-bit watermark of the $64 \times 64$ recovered block and the original one.
2. *Inside the top-left $32 \times 32$ subarea*: A pixel located at (270,199) with a gray value of 51 is randomly chosen. Let the value 51 be changed to 52. Figure 10.6c shows the modified watermarked *Lena*. The corresponding detection and localization result is shown in Figure 10.6d. Figure 10.7b shows the differences between the new 128-bit watermark of the $64 \times 64$ recovered block and the original one.

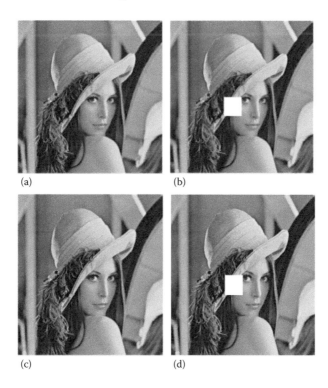

(a)                    (b)

(c)                    (d)

**FIGURE 10.6** Modification and localization in the sample block with its top-left corner position as (257,193): (a) one single LSB modification outside the top-left subarea, (b) the detection and localization result, (c) one single LSB modification within the top-left subarea, (d) the detection and localization result.

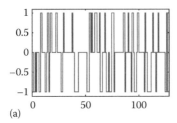 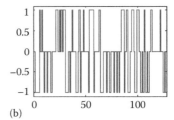

(a)                                      (b)

**FIGURE 10.7**   The differences between the new computed 128-bit watermark and the extracted 128-bit original in the sample block with its top-left corner position as (257,193): (a) modification outside the top-left subarea, (b) modification within the top-left subarea.

#### 10.1.2.4  Security

We use two keys in the proposed watermarking scheme: the initial condition $X_0 \in$ [0,1] and the initial parameter $P_0 \in$ (0,1) of the PWLCM. Under the control of corresponding keys, the chaotic hash value of each block of size $64 \times 64$ in the original and recovered images can be correctly computed.

On one hand, during the hashing process in the proposed scheme, not only is the extreme sensitivity to tiny changes in initial conditions and parameters of chaos fully utilized, so too is the mechanism of changeable parameters. This means the entire scheme possesses strong one-way properties and has nonlinear and sensitive dependence on the gray values, secret keys, and hash values. Therefore, it is computationally infeasible for someone to recover the keys. Besides, it is almost impossible to compute the hash value correctly without knowing the keys. Furthermore, the proposed scheme is immune from vector quantization and oracle attacks [2,3,23], since different gray values or secret keys will lead to different watermarks, even if the differences are very few.

On the other hand, the key-space size of the proposed scheme is large enough to resist brute-force attacks. For instance, a sample block of size $64 \times 64$ with its top-left corner as (257,193) in the watermarked *Lena* is used to demonstrate this. Let a very tiny change on the initial value $X_0$ of the PWLCM be $10^{-16}$; for example, $X_0$ is changed from 0.232323 to 0.2323230000000001. The corresponding changed bit number of the 128-bit watermark is around 64. Similarly, let a very tiny change on the initial parameters $P_0$ of the PWLCM be $10^{-15}$; for example, $P_0$ is changed from 0.858485 to 0.858485000000001. Figures 10.8a,b show the 128-bit watermark differences under slightly different $X_0$ and $P_0$ values, respectively. If the tiny changes to $X_0$ and $P_0$ are set as $10^{-17}$ and $10^{-16}$, respectively, still no corresponding bit of the watermark value changes.

We repeat this for all blocks of size $64 \times 64$. Figure 10.9 demonstrates the changed 128-bit watermark bit number in all the 64 blocks under slightly different $X_0$ and $P_0$ values. They are close to the ideal 64 bits, which is a half of the total 128 bits, indicating the dependence of the watermark on the secret key.

#### 10.1.2.5  Implementation and Other Ideas for Improvement

In the proposed scheme, double-precision floating-point arithmetic is involved. As long as the IEEE 754 standard is used, two watermark values of an image with the

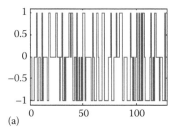 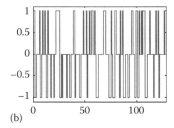

**FIGURE 10.8** The 128-bit watermark differences in the sample block with its top-left corner position as (257,193): (a) under slightly different $X_0$ values, (b) under slightly different $P_0$ values.

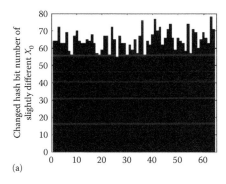 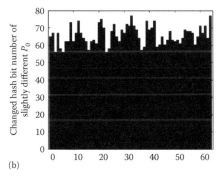

**FIGURE 10.9** The changed watermark bit number in all the 64 blocks under slightly different keys: (a) under slightly different $X_0$ values, (b) under slightly different $P_0$ values.

same secret key produced on two different computing platforms will be the same. Therefore, our scheme is practical in spite of using different operating systems and different program languages.

However, although the proposed scheme has reliable abilities in terms of content authentication and modification localization, it cannot recover from malicious modification when modification has been found. Our future work will involve finding more suitable solutions to overcome it—for example, by embedding some approximation of an image into a certain part of the image itself.

## 10.2   REVERSIBLE DATA-HIDING TECHNIQUES USING MULTIPLE-SCAN DIFFERENCE-VALUE HISTOGRAM MODIFICATION

We present an improved reversible data-hiding algorithm using multiple scanning techniques and histogram modification. Two performance metrics are important to consider with a reversible data-hiding technique. The first is the maximum payload, which is the maximum amount of data that can be embedded into an image. The second is visual quality, which is measured using the PSNR between the original image and the embedded image.

This section is organized as follows: In Section 10.2.1, we provide an overview of the proposed multiple scanning histogram modification algorithm. The single scanning histogram modification algorithm and multiple scanning histogram modification algorithms are described in Sections 10.2.2 and 10.2.3, respectively. Section 10.2.4 presents an algorithm for determining the embedding level and scan order. Section 10.2.5 shows experimental results. Conclusions are drawn in Section 10.2.6.

### 10.2.1   OVERVIEW OF THE MULTIPLE-SCAN HISTOGRAM MODIFICATION ALGORITHM

Our proposed algorithm consists of two stages: encoding the payload within an image to produce the cover image and decoding the cover image to retrieve the payload and the original image. We use three different scanning techniques for the construction of the difference histograms in order to maximize the payload capacity. The scanning techniques that we use are horizontal, vertical, and diagonal. The horizontal scanning technique is the same as that used in Zhao et al.'s algorithm [24]. Figure 10.10 shows the scanning order on a $4 \times 4$ image. The scanning order uses adjacent pixels since these are more likely to have similar values and will result in a difference histogram that has most of its values near zero.

Let us introduce the following symbols and notations that will be used throughout this chapter.

EL:   The embedding level, an integer parameter
$I$:     The original image
$I_L$:    An image that has been embedded L times
L:     The layer of embedding—that is, the number of times the histogram-shifting algorithm has been applied to the image
LM:   The location map of pixels that can be used in data embedding
$LM_C$: The compressed location map
P:     The payload embedded in the image

*Embedding process of our reversible data-hiding algorithm*:

1. Determine histogram shift order based on the payload.
2. Pairs of values corresponding to the scanning method and EL are outputted (discussed in detail in Section 10.2.5). The EL controls the hiding capacity (EL ≥ 0). A larger EL reflects that more secret data can be embedded.

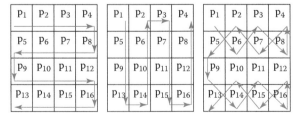

**FIGURE 10.10**   Pixel scan order for a $4 \times 4$ image.

3. Calculate the minimum number of pixels required to embed $LM_C$ .
4. Segment the image into $I_1$ and $I_2$.
5. Calculate LM of $I_1$.
6. Extract the LSBs of $I_2$ and concatenate them with the payload.
7. Compress LM using arithmetic coding and embed it within the LSB of the pixels of $I_2$.
8. Use the given scanning method to calculate the difference histogram of $I_1$ minus the bits in LM. Concatenate the key, $I_2$, and P to make $P_M$.
9. Shift the difference histogram for the given EL and embed $P_M$.
10. Apply the new difference histogram to the image.
11. Repeat steps 3–10 until all of the histogram shifts are applied and the payload is embedded. Save the scanning method, EL, and the key of the last histogram shift as external data.

*Extraction process of our reversible data-hiding algorithm*:

1. Use the scanning method, EL, and key provided.
2. Extract $LM_C$ from $I_L$ and decode into LM.
3. Calculate the embedded difference histogram of $I_L$.
4. Recover the original difference histogram and extract $P_M$.
5. Separate $P_M$ into the LSB of $I_2$, the next decoding method, EL, the key (if applicable), and P.
6. Use the original difference histogram to calculate $I_{L-1}$.
7. If $P_M$ contains a decoding method, EL, and key, use these values and go to step 2.
8. Concatenate all of the bits of P.

## 10.2.2   SINGLE-SCAN HISTOGRAM MODIFICATION

Zhao et al. [24] proposed a reversible data-hiding algorithm using a single-scan technique to produce the difference histogram. The scanning method used is the *inverse S*, which is the same as the horizontal scanning order in our algorithm. Since one scanning technique is used, a single difference histogram is produced. Figure 10.11 shows the *Lena* image and its histogram, with difference values between −20 and 20 using inverse S scanning.

The difference histogram of the *Lena* image clearly shows that most differences are around zero, with values farther from zero becoming less frequent. Using this characteristic, Zhao et al. uses pixels with difference values close to zero for data embedding. Data are embedded by shifting the difference histogram, such that two spaces are available for each difference value that has data embedded. This is to accommodate the binary values 0 and 1. Figure 10.12 shows the shifted *Lena* difference-value histogram and after data are embedded at each level.

The maximum capacity and PSNR of Zhao et al.'s algorithm depends on the EL and, consequently, the number of histogram shifts that are made to the difference values. Since the EL can be increased such that all pixels are used for data embedding, the theoretical maximum of an image's data-embedding capacity is 1

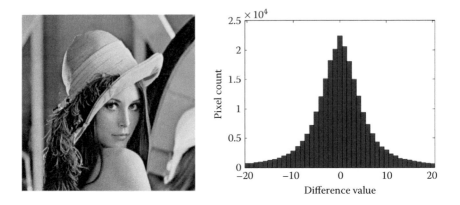

**FIGURE 10.11**   *Lena* image and the difference-value histogram using horizontal scanning.

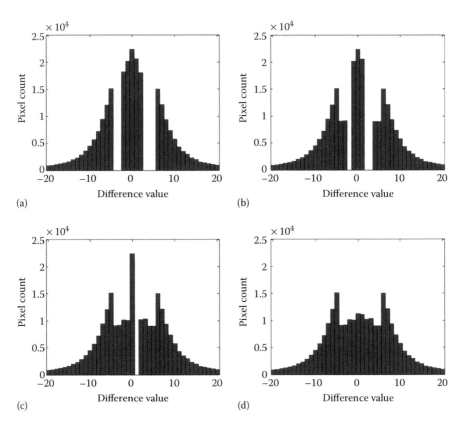

**FIGURE 10.12**   (a) Shifted *Lena* difference histogram and embedded histogram at (b) EL = 2, (c) EL = 1, (d) EL = 0.

bit per pixel (BPP). However, this maximum cannot be achieved using this technique because overhead information is required and some pixels may not be embeddable.

Since the data-embedding process can result in overflow (values >255) or underflow (values <0), the pixels whose values can take on these values after embedding cannot be altered. The pixels that are vulnerable to overflow or underflow are those in the intervals [0, EL] and [255 − EL, 255]. Not only must these pixels remain unchanged, but their locations must be saved and relayed to the decoder, so they are skipped during the decoding process as well. Zhao et al. used an M × N binary location map LM to prevent overflow and underflow. Before data are embedded, all pixels that are outside the valid range are excluded and have a value of 1 recorded in the LM; otherwise, a value of 0 is recorded. The LM is then compressed using arithmetic coding. Since the LM must be known to the decoder before the difference values are calculated, the compressed LM must be stored separately in the image. The image is split into two parts, $I_1$ and $I_2$, where $I_1$ is used to embed the payload and $I_2$ is used to store $LM_C$. $LM_C$ is stored in $I_2$ by replacing the LSBs of $I_2$ with $LM_C$. The LSBs of $I_2$ are then concatenated with the payload and the difference histogram-shifting algorithm is performed on the pixels of $I_1$.

The maximum data-embedding capacity of Zhao et al.'s algorithm is limited by two factors: the number of pixels available at each EL and the amount of data overhead caused by potential overflow and underflow. With an increasing EL, the number of pixels that are ineligible for embedding increases as well as the size of $LM_C$, which reduces the capacity for payload bits at each EL. In addition, the difference-value histograms at higher ELs have shorter peaks. These two factors result in a peak embedding capacity for images, as shown for the *Lena* image in Figure 10.13.

Figure 10.13 indicates that the embedding capacity of the single-scan method increases rapidly with a small number of histogram shifts and has greatly diminishing returns for higher histogram shifts. The embedding capacity gradually levels off

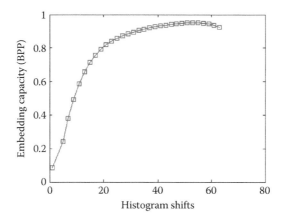

**FIGURE 10.13**    The embedding capacity of the *Lena* image using Zhao et al.'s algorithm.

and declines, which is when the overhead information is greater than the improvement in payload capacity. As each pixel can only hold 1 bit of payload data, Zhao et al.'s algorithm has a hard upper bound of 1 BPP. This limitation inspired our proposed technique, which attempts to break this capacity ceiling while meeting or exceeding the cover image quality at a given capacity.

### 10.2.3 MULTIPLE-SCAN HISTOGRAM MODIFICATION EXAMPLE

In this section we discuss the improvements made by using multiple scanning techniques in calculating the difference-value histogram. Figure 10.14 shows difference-value histograms of the *Lena* image using the horizontal, vertical, and diagonal scanning techniques. For the *Lena* image, the vertical scanning technique has higher peaks for the difference values close to zero. Therefore, we first perform histogram modification using the vertical scanning technique. For demonstrative purposes, we will arbitrarily choose an EL of 3 for embedding.

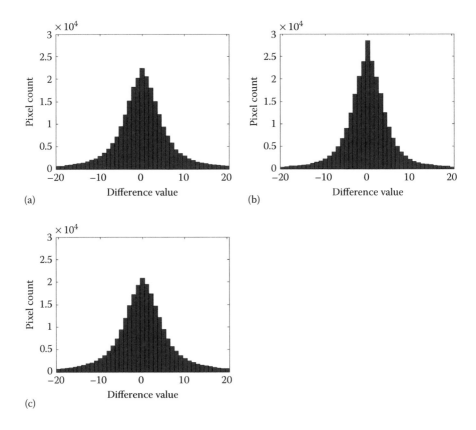

**FIGURE 10.14**  Difference-value histogram of the *Lena* image using (a) horizontal scanning, (b) vertical scanning, (c) diagonal scanning.

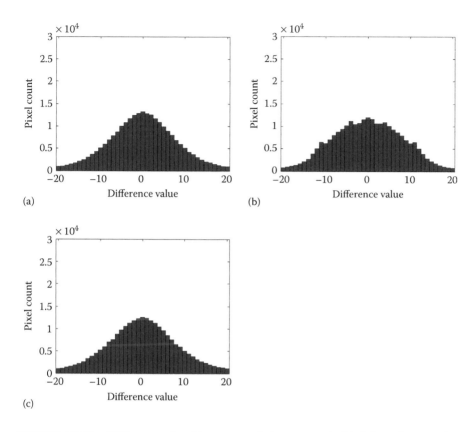

**FIGURE 10.15** Difference-value histograms of the once-embedded *Lena* image using (a) horizontal scanning, (b) vertical scanning, (c) diagonal scanning.

Figure 10.15 shows the *Lena* difference-value histograms after performing the horizontal shifting algorithm with an EL of 3. While all of the difference-value histogram peaks around zero have decreased, the peak values using the horizontal scanning technique are higher than those of the vertical scanning difference histogram of the original image for an EL greater than 3. As a result, the payload capacity is increased by first embedding using vertical scanning and then horizontal scanning rather than embedding using only one method at an increased EL.

We perform histogram shifting again, this time using horizontal scanning with an EL of 3 followed by diagonal scanning with an EL of 3. At this point we have embedded 303,076 bits of data for a payload capacity of 1.16 BPP. This exceeds the theoretical maximum payload of Zhao et al.'s algorithm. However, it is unclear by solely examining the difference-value histograms which order of scanning techniques should be chosen and what EL should be used. The next section presents an algorithm that provides a high-capacity embedding order.

## 10.2.4  Iterative Algorithm for Determining Embedding Level and Scan Order

A typical image can have over 40 histogram-shifting operations applied to it before a reduction in payload capacity is experienced. This means that there are approximately $40^3 = 64{,}000$ combinations of histogram shifts that can be applied to the image before the payload capacity is reduced. Since it is computationally expensive to perform such a large number of histogram shifts, a brute-force method of determining the histogram shift and EL order is undesirable. Therefore, we propose an iterative algorithm that aims to maximize the payload capacity while maintaining low computational complexity.

The proposed algorithm for optimizing the histogram shift and EL order is an application of the hill-climbing technique. We use this technique to maximize the payload capacity that is gained at each step while minimizing the increase in the cover image noise. We store a list of proposed methods. A proposed method is defined as a sequence of pairs of scanning methods and ELs. For each proposed method, we store the capacity and PSNR of the produced cover image at each EL.

We begin by initializing the proposed methods as the three scanning methods at EL = 0. The payloads and PSNRs of the produced cover images are saved with the corresponding method. The active method is defined as the method with the highest payload-to-PSNR ratio. The first active method is set as the maximum payload of the proposed methods. Then we perform the following iterative operation.

1. Apply each histogram shift using EL = 0 to the cover images of the current method using each scanning technique. We call the method used in this step the *current embedding method*.
    a. Compare the payload capacity and PSNR with the proposed method.
    b. If the payload capacity of a proposed method is less than that of the current embedding method, then increase the EL of the proposed method's last scanning technique until it is greater than that of the current embedding method or until the proposed method technique decreases.
    c. If the current embedding method has a greater payload capacity than all of the proposed methods for the PSNR of the cover image of the current embedding method, then the next step is defined as the current embedding method. Otherwise, the next step is defined as the proposed method with the highest payload capacity
    d. Repeat steps 2a–c for each of the scanning techniques.
3. If the next step is one of the proposed methods, then set the current method to the proposed method, changing the EL of the last step to the EL that was set in step 2b. Otherwise, append the next method to the current method.
4. Repeat steps 1–3 until the desired payload capacity is reached or there is no increase in the payload capacity.

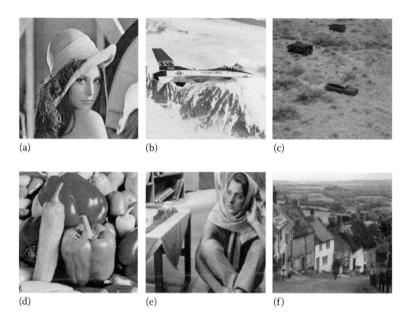

(a)                          (b)                          (c)

(d)                          (e)                          (f)

**FIGURE 10.16**    Test images: (a) *Lena*, (b) *Airplane*, (c) *Car*, (d) *Peppers*, (e), *Barbara*, (f) *Goldhill*.

### 10.2.5    EXPERIMENTAL RESULTS

We perform our proposed algorithm on six test images, as shown in Figure 10.16. Our proposed algorithm is compared with the single–scanning method algorithm. Payloads are pseudorandomly generated such that the maximum number of bits is embedded in each image. The PSNR and payload capacity are compared for each algorithm.

Figure 10.17 shows the payload capacity of the test images and the PSNR of the produced cover images using Zhao et al.'s algorithm and our proposed algorithm. Since the single horizontal scanning method is always in the solution scanning space of our EL and scan order algorithm, our algorithm always produces at least as much storage capacity for a given PSNR value. The proposed algorithm has a substantially improved maximum payload capacity. While Zhao et al.'s algorithm has a theoretical maximum of 1 BPP, our algorithm consistently exceeds this boundary, reaching higher than 2.5 BPP in the *Airplane* image.

Our algorithm has a higher maximum capacity at the 30 dB PSNR level as well as a higher overall maximum payload capacity for every test image. At the 30 dB PSNR level, which is considered to be the limit of human perception, our algorithm has an average maximum payload capacity of 1.1179 BPP, compared with the 0.8530 BPP of Zhao et al.'s algorithm, a 31% improvement. The overall average maximum payload capacity of our algorithm is 1.9233 BPP, compared with 0.8706 BPP, a 120% improvement. Table 10.2 shows the comparison of payload capacity.

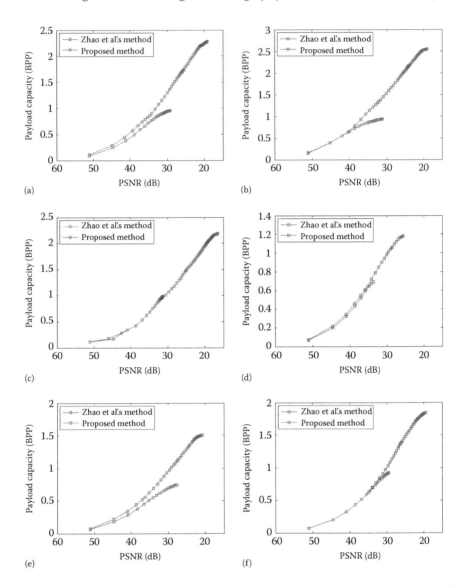

**FIGURE 10.17**   The noise of the cover images (PSNR) plotted with the payload capacity of Zhao's algorithm and the proposed algorithm of the following images: (a) *Lena*, (b) *Airplane*, (c) *Car*, (d) *Peppers*, (e), *Barbara*, (f) *Goldhill*.

## TABLE 10.2
## Comparison of Payload Capacities

|  | Zhao et al.'s Method | | Our Proposed Method | |
|---|---|---|---|---|
| Image | Max. Capacity PSNR <30 (BPP) | Max. Capacity Overall (BPP) | Max. Capacity PSNR <30 (BPP) | Max. Capacity Overall (BPP) |
| *Lena* | 0.9367 | 0.9516 | 1.2952 | 2.2786 |
| *Airplane* | 0.9344 | 0.9344 | 1.5227 | 2.5444 |
| *Car* | 0.9806 | 0.9806 | 1.0098 | 2.1889 |
| *Peppers* | 0.6903 | 0.6903 | 0.9805 | 1.1800 |
| *Barbara* | 0.6794 | 0.7478 | 0.9296 | 1.5062 |
| *Goldhill* | 0.8964 | 0.9190 | 0.9694 | 1.8416 |
| Average | 0.8530 | 0.8706 | 1.1179 | 1.9233 |

# REFERENCES

1. Yeung, M. and Mintzer, F., An invisible watermarking technique for image verification, in *Proc. IEEE Int. Conf. Image Processing*, Santa Barbara, CA, 680, 1997.
2. Fridrich, J., Security of fragile authentication watermarks with localization, in *Proc. SPIE Conf. Security and Watermarking of Multimedia Contents*, San Jose, CA, 691, 2002.
3. Holliman, M. and Memon, N., Counterfeiting attacks on oblivious block-wise independent invisible watermarking schemes, *IEEE Trans. Image Processing*, 9, 432, 2000.
4. Celik, M. U., Sharma, G., Saber, E., and Tekalp, A. M., Hierarchical watermarking for secure image authentication with localization, *IEEE Trans. Image Processing*, 11, 585, 2002.
5. Chan, C. S. and Chang, C. C., An efficient image authentication method based on Hamming code, *Pattern Recognition*, 40, 681, 2007.
6. Liu, S. H., Yao, H. X., Gao, W., and Liu, Y. L., An image fragile watermark scheme based on chaotic image pattern and pixel-pairs, *Applied Mathematics and Computation*, 185, 869, 2007.
7. Zhang, X. P. and Wang, S. Z., Fragile watermarking scheme using a hierarchical mechanism, *Signal Processing*, 89, 675, 2009.
8. Chen, C. C. and Lin, C. S., A GA-based nearly optimal image authentication approach, *Int. J. Innovative Computing, Information and Control*, 3, 631, 2007.
9. Luo, X. Y., Liu, F. L., and Liu, P. Z., An LSB steganography approach against pixels sample pairs steganalysis, *Int. J. Innovative Computing, Information and Control*, 3, 575, 2007.
10. Tian, J., Reversible data embedding using a difference expansion, *IEEE Trans. Circuits Systems for Video Technology*, 13, 890, 2003.
11. Alattar, A. M., Reversible watermark using the difference expansion of a generalized integer transform, *IEEE Trans. Image Processing*, 13, 1147, 2004.
12. Kim, H. J., Sachnev, V., Shi, Y. Q., Nam, J., and Choo, H. G., A novel difference expansion transform for reversible data embedding, *IEEE Trans. Information Forensics and Security*, 3, 456, 2008.
13. Weng, S. W., Zhao, Y., and Pan, J. S., A novel reversible data hiding scheme, *Int. J. Innovative Computing, Information and Control*, 4, 351, 2008.

14. Ni, Z., Shi, Y. Q., Ansari, N., and Su, W., Reversible data hiding, *IEEE Trans. Circuits and Systems for Video Technology*, 16, 354, 2006.
15. Lin, C. C., Tai, W. L., and Chang, C. C., Multilevel reversible data hiding based on histogram modification of difference images, *Pattern Recognition*, 41, 3582, 2008.
16. Chen, C. C. and Kao, D. S., DCT-based zero replacement reversible image watermarking approach, *Int. J. Innovative Computing, Information and Control*, 4, 3027, 2008.
17. Coltuc, D. and Chassery, J. M., Very fast watermarking by reversible contrast mapping, *IEEE Signal Processing Letters*, 14, 255, 2007.
18. Sudhamani, M. V. and Venugopal, C. R., Nonparametric classification of data viz. clustering for extracting color features: An application for image retrieval, *ICIC Express Letters*, 1, 15, 2007.
19. Xiao, D., Liao, X. F., and Deng, S. J., One-way Hash function construction based on the chaotic map with changeable-parameter, *Chaos, Solitons and Fractals*, 24, 65, 2005.
20. Xiao, D., Liao, X. F., and Deng, S. J., Parallel keyed hash function construction based on chaotic maps, *Physics Letters A*, 372, 4682, 2008.
21. Yi, X., Hash function based on the chaotic tent map, *IEEE Trans. Circuits and Systems II*, 52, 354. 2005.
22. Zhang, J. S., Wang, X. M., and Zhang, W. F., Chaotic keyed hash function based on feedforward–feedback nonlinear digital filter, *Physics Letters A*, 362, 439, 2007.
23. Wu, J. H., Zhu, B. B., Li, S. P., and Lin, F. Z., New attacks on sari image authentication system, in *Proc. SPIE*, San Jose, CA, 602, 2004.
24. Zhao, Z., Luo, H., Lu, Z., and Pan, J., Reversible data hiding based on multilevel histogram modification and sequential recovery, *AEU: Int. J. Electronics and Communications*, 65, 814, 2011.

# 11 Steganography and Steganalysis

As digital information and data are transmitted more often than ever over the Internet, the technology of protecting and securing sensitive messages needs to be discovered and developed. Digital steganography is the art and science of hiding information in covert channels, so as to conceal the information and prevent the detection of hidden messages. Covert channel communication is a genre of information security research that generally is not a major player but has been an extremely active topic in the academic, industrial, and governmental domains for the past 30 years.

Steganography and cryptography are used in data hiding. Cryptography is the science of protecting data by scrambling, so that nobody can read them without given methods or keys. It allows an individual to encrypt data so that the recipient is the only person able to decipher them. Steganography is the science of obscuring a message in a *carrier* (host object) with the intent of not drawing suspicion to the context in which the messages was transferred.

The term *steganography* is derived from the Greek words *steganos*, meaning "covered," and *graphein*, meaning "to write." The intention is to hide information in a medium in such a manner that no one except the anticipated recipient knows of its existence. As children, we might have experienced writing an invisible message in lemon juice and asking people to find out the secret message by holding it against a light bulb. Hence, we may already have used the technology of steganography in the past.

The history of steganography can be traced back to around 440 BC, where the Greek historian Herodotus described two events in his writings [1], one involving wax to cover secret messages, and the other involving shaved heads. In ancient China, military generals and diplomats hid secret messages on thin sheets of silk or paper. One famous story of the successful revolt of the Han Chinese against the Mongolians during the Yuan dynasty involves a steganographic technique. During the Mid-Autumn Festival, the Han people inserted messages (as hidden information) inside mooncakes (as cover objects) and gave these cakes to their members to deliver information on the planned revolt. Nowadays, with the advent of modern computer technology, a great amount of state-of-the-art steganographic algorithms have used old-fashioned steganographic techniques to reform the literature on information hiding, watermarking, and steganography.

Modern techniques in steganography are far more powerful. Many software tools allow a paranoid sender to embed messages in digitized information, typically audio, video, or still image files that are sent to a recipient. Although steganography has attracted great interest from the military and governmental organizations, there is even a lot of interest from commercial companies wanting to safeguard their information from piracy. Today, steganography is often used to transmit information safely and embed trademarks securely in images and music files to assure copyright.

The goal of steganography is to avoid drawing suspicion to the transmission of a secret message. Its essential principle requires that the stego-object (i.e., the object containing hidden messages) should be perceptually indistinguishable to the degree that suspicion is not raised. Detecting the secret message, called *steganalysis*, relies on the fact that hiding information in digital media alters the characteristics of the carrier and applies some sort of distortion to it. By analyzing the various features of stego-images and cover images (i.e., the images containing no hidden messages), a steganalytic system is developed to detect hidden data.

## 11.1 STEGANOGRAPHY

In this section, we introduce the types of steganography, its applications, embedding, security and imperceptibility, and some examples of steganographic software.

### 11.1.1 TYPES OF STEGANOGRAPHY

Steganography can be divided into three types: technical, linguistic, and digital. Technical steganography applies scientific methods to conceal secret messages, while linguistic steganography uses written natural language. Digital steganography, developed with the advent of computers, employs computer files or digital multimedia data. In this section, we describe each type respectively.

#### 11.1.1.1 Technical Steganography

Technical steganography applies scientific methods to conceal a secret message, such as the use of invisible ink, microdots, and shaved heads. Invisible ink is the simplest method of writing secret information on a piece of paper, so that when it dries, the written information disappears. The paper looks the same as the original blank piece of paper. We can use organic compounds such as milk, urine, vinegar, or fruit juice. When we apply heat or ultraviolet light, the dry message will become visible to human perception. In 1641, Bishop John Wilkins invented invisible ink using onion juice, alum, and ammonia salts. Modern invisible inks fluoresce under ultraviolet light and are used as counterfeit-detecting devices.

With the advent of photography, microfilm was created as a medium of recording a great amount of information in a very small space. In World War II, the Germans used microdots to convey secret information. The technique is famous as "the enemy's masterpiece of espionage," as J. Edgar Hoover called it. The secret message was first photographed and then reduced to the size of a printed period to be pasted onto innocuous host documents such as newspapers or magazines. Microdots are printed as many small dots. Because their sizes are so small, people do not pay attention to them. However, with the transmission of enormous amounts of microdots, one can hide secret messages, drawings, or even images. Another kind is to use printers to print great amounts of small yellow dots, so they are hardly visible under normal white-light illumination. The pattern of small yellow dots can construct secret messages. When we place it under different color light, for example, blue light, the pattern of yellow dots becomes visible.

Steganography using shaved heads started when Histaeus, the ruler of Miletus, planned to send a message to his friend Aristagorus, urging his revolt against the Persians. He shaved the head of his most trusted slave as the messenger and then tattooed a message or a symbol on the messenger's head. After the hair grew back, the messenger ran to Aristagorus to deliver the hidden message. When he arrived, his head was shaved again to reveal the secret message.

Later in Herodotus's history, the Spartans received a secret message from Demeratus that Xerxes of the Persian army was preparing to invade Greece. Demeratus was a Greek in exile in Persia. Fearing discovery, he securely concealed the secret message by scraping the wax off a writing tablet, inscribing the message on the wood, and then re-covering the tablet with wax to make it look normal. This tablet was delivered to the Spartans, who, upon receiving the message, hastened to rescue Greece.

### 11.1.1.2 Linguistic Steganography

Linguistic steganography utilizes written natural language to hide information. It can be categorized into *semagrams* and *open codes*.

Semagrams hide secret messages using visual symbols or signs. Modern updates to this method use computers to make the secret message even less noticeable. For example, we can alter the font size of specific letters or make their locations a little bit higher or lower to conceal messages. We can also add extra spaces in specific locations of the text or adjust the spacing of lines. Semagrams can also be encoded into pictures. The most well-known form of this is the secret message in *The Adventure of the Dancing Men* by Sir Arthur Conan Doyle.

Semagrams as described above are sometimes not secure enough. Another secret method is used when spies want to set up meetings or pass information to their networks. This includes hiding information in everyday things, such as newspapers, dates, clothing, or conversation. Sometimes meeting times can be hidden in reading materials. In one case, an advertisement for a car was placed in a city newspaper during a specific week that read, "Toyota Camry, 1996, needs engine work, sale for $1500." If the spy saw this advertisement, he/she knew that the Japanese contact wanted to have personal communication immediately.

Open codes hide secret messages in a document in a specifically designed pattern that is not obvious to the average reader. The following example of an open code was used by a German spy in World War II.

Apparently neutral's protest is thoroughly discounted and ignored. Isman hard hit. Blockade issue affects pretext for embargo on by-products, ejecting suets and vegetable oils.

If we collect the second letter in each word, we can extract the secret message as

**Pershing sails from NY June 1.**

### 11.1.1.3   Digital Steganography

Computer technology has made it a lot easier to hide messages and more difficult to discover them. Digital steganography is the science of hiding secret messages within digital media, such as digital images, audio files, or video files. There are many different methods for digital steganography, including least-significant-bit (LSB) substitution, message scattering, masking and filtering, and image-processing functions.

Secret information or images can be hidden in image files because image files are often very large. When a picture is scanned, a sampling process is conducted to quantize the picture into a discrete set of real numbers. These samples are the gray levels in an equally spaced array of pixels. The pixels are quantized to a set of discrete gray-level values, which are also taken to be equally spaced. The result of sampling and quantizing produces a digital image. For example, if we use 8-bit quantization (i.e., 256 gray levels) and 500-line squared sampling. This would generate an array of 250,000 8-bit numbers. It means that a digital image will need 2 million bits.

We can also hide secret messages on a hard drive in a secret partition. The hidden partition is invisible, although it can be accessed by disk configuration and other tools. Another form is to use network protocols. By using a secret protocol that includes the sequence number field in transmission control segments and the identification field in Internet protocol packets, we can construct covert communication channels.

Modern digital steganographic software employs sophisticated algorithms to conceal secret messages. It represents a particularly significant threat nowadays, due to the huge amount of digital steganography tools freely available on the Internet that can be used to hide any digital file inside of another. With easy access and simple usage, criminals are inclined to conceal their activities in cyberspace. It has been reported that even Al Qaida terrorists use digital steganography tools to deliver messages. With the devastating attack on the World Trade Center in New York City on September 11, 2001, there were indications that the terrorists had used steganography to conceal their correspondence during the planning of the attack. Therefore, digital steganography presents a grand challenge to law enforcement as well as industry, because detecting and extracting hidden information is very difficult.

### 11.1.2   Applications of Steganography

### 11.1.2.1   Convert Communication

A classic steganographic model is the prisoners' problem presented by Simmons [2], although it was originally introduced to describe a cryptography scenario. Two inmates called Alice and Bob are locked in separate jail cells and intend to communicate a secret plan to escape together. All communications between them are monitored by Wendy, the warden. Alice and Bob know that Wendy will terminate their communications channels if she discovers something strange.

Fitting into the steganography model, Alice and Bob must hide the escape messages in other innocuous-looking media, called *cover objects*, to construct stego-objects for Wendy's inspection. Then, the stego-objects are sent through the public channel for communication. Wendy is free to inspect all the messages between Alice and Bob in two ways: *passively* or *actively*. The passive way is to inspect the stego-object carefully in order to determine whether it contains a hidden message and then take the proper action. On the other hand, the active way is to always alter the stego-object to destroy hidden messages, although Wendy may not perceive any trace of one.

The basic applications of steganography relate to secret communications. Modern computer and networking technology allows individuals, groups, and companies to host a Web page that may contain secret information meant for another party. Anyone can download the Web page; however, the hidden information is invisible and does not draw any attention. The extraction of secret information requires specific software with the correct keys. Adding encryption to the secret information will further enhance its security. The situation is similar to hiding some important documents or valuable merchandise in a secure safe. Furthermore, we hide the safe in a secret place that is difficult to find.

All digital data files can be used for steganography, but the files containing a superior degree of *redundancy* are more appropriate. Redundancy is defined as the number of bits required for an object to provide necessarily accurate representation. If we remove the redundant bits, the object will look the same. Digital images and audio mostly contain a great amount of redundant bits. Therefore, they are often used as cover objects.

### 11.1.2.2   One-Time Pad Communication

The *one-time pad* was originated in cryptography to provide a random private key that can only be used once for encrypting a message; we then can decrypt it using a one-time matching pad and key. It is generated using a string of numbers or characters having the length at least as long as the longest message sent. A random number generator is used to generate a string of values randomly. These values are stored on a pad or a device for someone to use. Then, the pads are delivered to the sender and the receiver. Usually, a collection of secret keys are delivered and can only be used once in such a manner that there is one key for each day in a month, and the key expires at the end of each day.

Messages encrypted by a randomly generated key possess the advantage that there is theoretically no way to uncover the code by analyzing a series of the messages. All the encryptions are different in nature and bear no relationship. This kind of encryption is referred to as a *100% noise source*. Only the sender and receiver are able to delete the noise. We should note that a one-time pad can be only used once, for security reasons. If it is reused, someone may perform comparisons of multiple messages to extract the key for deciphering the messages. One-time pad technology was adopted notably in secret communication during World War II and the Cold War. In today's Internet communication, it is also used in public-key cryptography.

The one-time pad is the only encryption method that can be mathematically proved to be unbreakable. The technique can be used to hide an image by splitting it into two random layers of dots. When they are superimposed, the image appears. We can generate two innocent-looking images in such a way that putting both images together reveals the secret message. The encrypted data or images are perceived as a set of random numbers. It is obvious that no warden will permit the prisoners to exchange encrypted, random, malicious-looking messages. Therefore, one-time pad encryption, though statistically secure, is practically impossible to use in this scenario.

### 11.1.3 Embedding Security and Imperceptibility

A robust steganographic system must be extremely secure from all kinds of attacks (i.e., steganalysis) and must not degrade the visual quality of the cover images. We can combine several strategies to ensure security and imperceptibility. For security we can use a *chaotic mechanism* (CM), a *frequency hopping* (FH) structure, a *pseudorandom number generator* (PNG), or a *patchwork locating algorithm* (PLA). For imperceptibility we can use *set partitioning in hierarchical trees* (SPIHT) coding, the *discrete wavelet transform* (DWT), and *parity check embedding* (PCE).

The benchmarking of steganographic techniques is a complicated task that requires the examination of a set of mutually dependent performance indices [3–9]. A benchmarking tool should be able to interact with different performance aspects, such as visual quality and robustness. The input to the benchmarking system should contain images that vary in size and frequency content, since these factors affect the system's performance.

We should also evaluate the execution time of the embedding and detection modules. Mean, maximum, and minimum execution times for the two modules might be evaluated over the set of all keys and messages for each host image. We should measure the perceptual quality of stego-images by subjective quality evaluation and in a quantitative way that correlates well with the way human observers perceive image quality. We should also evaluate the maximum number of information bits that can be embedded per host image pixel.

Since steganography is expected to be robust to host image manipulation, tests for judging the performance of a steganographic technique when applied to distorted images constitute an important part of a benchmarking system. In order to make the embedded messages undetectable, the set of attacks available in a benchmarking system should include all operations that the average user or an intelligent pirate can use. It should also include signal-processing operations and distortions that occur during normal image usage, transmission, storage, and so on.

### 11.1.4 Examples of Steganographic Software

Steganography embeds secret messages into innocuous files. Its goal is not only to hide the message but to also let the stego-image pass without causing suspicion. There are many steganographic tools available for hiding messages in image, audio, and video files. In this section, we briefly introduce S-Tools, StegoDos, EzStego, and JSteg-Jpeg.

#### 11.1.4.1  S-Tools

The S-Tools program [10], short for Steganography Tools, was written by Andy Brown to hide secret messages in BMP, GIF, and WAV files. It is a combined steganographic and cryptographic product, since the messages to be hidden are encrypted using symmetric keys. It uses LSB substitution in files that employ lossless compression, such as 8- or 24-bit color- and pulse-code modulation, and applies a pseudorandom number generator to make the extraction of secret messages more difficult. It also provides the encryption and decryption of hidden files with several different encryption algorithms.

We can open up a copy of S-Tools and drag images and audio across to it. To hide files we can drag them over open audio or image windows. We can hide multiple files in one audio or image, and compress the data before encryption. The multithreaded functionality provides the flexibility of many hide/reveal operations running simultaneously without interfering with other work. We can also close the original image or audio without affecting the ongoing threads.

### 11.1.4.2    StegoDos

StegoDos [11], also known as Black Wolf's Picture Encoder, consists of a set of programs that allow us to capture an image, encode a secret message, and display the stego-image. This stego-image may also be captured again into another format using a third-party program. Then we can recapture it and decode the hidden message. It only works with $320 \times 200$ images with 256 colors. It also uses LSB substitution to hide messages.

### 11.1.4.3    EzStego

EzStego, developed by Romana Machado, simulates invisible ink in Internet communications. It hides an encrypted message in a GIF-format image file by modulating the LSBs. It begins by sorting the colors in the palette so that the closest colors fall next to each other. Then similar colors are paired up, and for each pair one color will represent 1 while the other will represent 0. It encodes the encrypted message by replacing the LSBs. EzStego compares the bits it wants to hide with the color of each pixel. If the pixel already represents the correct bit, it remains untouched. If the pixel represents the incorrect bit, the color would be changed to its pair color. The changes to the LSBs are so small that they are undetectable by human perception. EzStego treats the colors as cities in the three-dimensional RGB space, and intends to search for the shortest path through all of the stops.

### 11.1.4.4    JSteg-Jpeg

JSteg-Jpeg, developed by Derek Upham, can read multiple image formats and embed secret messages to be saved as JPEG format images [12]. It utilizes the splitting of the JPEG encoding into lossy and lossless stages. The lossy stages use the *discrete cosine transform* (DCT) and a quantization step to compress the image data, and the lossless stage uses Huffmann coding to compress the image data. The DCT is a compression technique since the cosine values cannot be accurately calculated and the repeated calculations can introduce rounding errors. It therefore inserts the secret message into the image data between these two steps. It also modulates the rounding processes in the quantized DCT coefficients. Image degradation is affected by the embedded amount as well as the quality factor used in the JPEG compression. The trade-off between the two can be adjusted to allow for imperceptibility.

## 11.2   STEGANALYSIS

Like digital watermarking, digital steganalytic systems can be categorized into two classes: *spatial domain steganalytic systems* (SDSSs) and *frequency domain steganalytic systems* (FDSSs). SDSSs [13,14] are adopted for checking the lossless compressed images by analyzing statistical features in the spatial domain. The simplest SDSS is LSB substitution. This technique considers the LSB to be random noise, so its modification will not affect the overall image visualization. For lossy compression images such as JPEGs, FDSSs [15,16] are used to analyze statistical features in the frequency domain. Westfeld and Pfitzmann [13] presented two SDSSs based on visual and chi-square attacks. Visual attacks use human sight to inspect stego-images by checking their lower bit planes. Chi-square attacks can automatically detect the specific characteristic generated by the LSB steganographic technique. Avcibas et al. [14] proposed the *image quality measure* (IQM), which is based on the hypothesis that the steganographic systems leave statistical evidence that can be exploited for detection using multivariate regression analysis. Fridrich et al. [15] presented an FDSS for detecting JPEG stego-images by analyzing their DCT with cropped images.

Steganalysis is the process of detecting steganography. Basically, there are two methods of detecting modified files. One is called *visual analysis*, which involves comparing a suspected file with the original copy. It intends to reveal the presence of secret communication through inspection, either by eyes or with the help of a computer system, typically decomposing the image into its bit planes. Although this method is very simple, it is not very effective; most of the time, the original copy is unavailable.

The other is called *statistical analysis*, which detects changes in the patterns of pixels and the frequency distribution of the intensities. It can detect whether an image has been modified by checking whether its statistical properties deviate from the norm. Therefore, it intends to find even slight alterations to the statistical behavior caused by steganographic embedding.

Steganalytic techniques must be updated frequently and evolve continuously in order to combat new and more sophisticated steganographic methods. Even the U.S. law enforcement agencies have no steganographic guideline available to them. However, there are certain rules that investigators adopt when they are looking for hints that may suggest the use of steganography. These include the technical capabilities of the computer's owner, software clues, program files, multimedia files, and the type of crime.

In this section, we introduce image statistical properties, visual steganalytic systems (VSSs), the IQM-based steganalytic system, learning strategies, and FDSSs.

### 11.2.1   Image Statistical Properties

There are statistical properties in existence for the human perception of natural images. The wavelength of light generates visual stimuli. When the wavelength dynamically changes, the perceived color varies from red, orange, yellow, green, and blue to violet. Human visual sensitivities are closely related to color; for example, humans are more sensitive to green than red and blue. We can divide the composition of a color into three components: brightness, hue, and saturation. The eye's adaptation to a particular hue distorts the perception of other hues; for example, gray seen after green, or seen against a green background, looks purplish.

Researchers in statistical steganalysis have put effort into advancing their methods due to the increasing preservation of first-order statistics in steganography to avoid detection. The encryption of hidden message also makes detection more difficult because the encrypted data generally have a high degree of randomness, and ones and zeros exist in equal likelihood [16,17]. In addition to detection, the recovery of hidden messages appears to be a complicated problem, since it requires knowledge of the crypto-algorithm or the encryption key [18].

LSB insertion in a palette-based image will produce a great amount of duplicate colors. Some identical (or nearly identical) colors may appear twice in the palette. Some steganographic techniques rearranging the color palette order for hiding messages may cause structural changes, which will generate the signature of the steganography algorithm [19,20].

Similar colors tend to be clustered together. Changing the LSB of a color pixel does not degrade the overall visual perception—for example, in LSB steganography. There are four other steganalytic methods, including the *regular–singular* (RS) method, *pairs of values* method, *pairs analysis* method, and *sample pair* method, which take advantage of statistical changes by modeling changes in certain values as a function of the percentage of pixels with embedded steganography. We briefly describe each method below.

The RS method was developed by Fridrich et al. [21]. They divide a cover image into a set of $2 \times 2$ disjoint groups instead of $4 \times 1$ slices. A discrimination function is applied to quantify the smoothness or regularity of the group of pixels. The variation is defined as the sum of the absolute value of the difference between adjacent pixels in the group. An invertible operation, called *shifted LSB flipping*, is defined as a permutation that entirely consists of two cycles.

The pairs of values method was developed by Pfitzman and Westfeld [22]. They use a fixed set of *pairs of values* (PoVs) to flip for embedding message bits. They construct the PoVs using quantized DCT coefficients, pixel values, or palette indices, and spread out two values from each pair irregularly in the cover image before message embedding. The occurrences of the values in each pair are inclined to be equal after embedding. The sum of the occurrences of a pair of colors in the image remains the same if we swap one value for another. This concept is used in designing a statistical chi-square test.

The pairs analysis method was developed by Fridrich et al. [23]. It can detect steganography with diluted changes uniformly straddled over the whole carrier medium. It can also estimate the embedded message length. The pairs of colors that are exchanged during LSB embedding correspond to a preordering of the palette by LSB steganography.

The sample pair method was developed by Dumitrescu et al. [24]. The image is represented as a set of sample pairs, which are divided into sub-multisets according to the difference between the pixel value and the value after LSB embedding. They assume that the probability of sample pairs having a large even component is equal to that having a large odd component for natural images. Therefore, the accuracy of their method depends on the accuracy of this assumption. They use state machines to statistically measure the cardinalities of the trace multisets before and after LSB embedding.

Steganographic techniques generally modify the statistical properties of the carrier; a longer hidden message will alter the carrier more than a shorter one [16,25–27]. Statistical analysis is utilized to detect hidden messages, particularly in blind steganography [20]. Statistical analysis of images can determine whether their properties deviate from the expected norm [16,17,27]. These include means, variances, and chi-square ($\chi^2$) tests [19].

## 11.2.2 Visual Steganalytic Systems

Kessler [28] presented an overview of steganography for computer forensics examiners. Steganalysis strategies usually follow the method that the steganography algorithm uses. A simple way is to visually inspect the carrier and steganography media. Many simple steganographic techniques embed secret messages in the spatial domain and select message bits in the carrier independently of its content [19].

Most steganographic systems embed message bits in a sequential or pseudorandom manner. If the image contains certain connected areas of uniform color or saturated color (i.e., 0 or 255), we can use visual inspection to find suspicious artifacts. If the artifacts are not found, an inspection of the LSB bit plane is conducted.

Westfeld and Pfitzmann [22] presented a VSS that uses an assignment function of color replacement called the *visual filter*. The idea is to remove all parts of the image that contain a potential message. The filtering process relies on the presumed steganographic tool and can attack the carrier medium steganogram, extract potential message bits, and produce a visual illustration.

## 11.2.3 IQM-Based Steganalytic Systems

Image quality can be characterized by the correlation between the display and the *human visual system* (HVS). With the advent of the IQM, we aim at minimizing the subjective decisions. Computational image quality analysis is geared toward developing an accurate computational model to describe the image quality. There are two types of IQM: full-referenced IQM and no-reference IQM. The full-referenced IQM has the original (i.e., undistorted) image available as a reference to be compared with the distorted image. It applies different kinds of distance metrics to measure the closeness between them. The no-reference IQM just takes the distorted image alone for analysis. It uses image-processing operators, such as edge detectors, to compute the quality index.

Avcibas et al. [29] developed a technique based on the IQM for the steganalysis of images that have been potentially subjected to steganographic algorithms, both within the passive warden and active warden frameworks. They assume that steganographic schemes leave statistical evidence that can be exploited for detection with the aid of image quality features and multivariate regression analysis. To this effect, image quality metrics have been identified, based on variance analysis techniques, as feature sets to distinguish between cover images and stego-images. The classifier between cover images and stego-images is built using multivariate regression on the selected quality metrics and is trained based on an estimate of the original image.

Generally, we can categorize IQMs into three types, as follows. Let $F(j,k)$ denote the pixel value in row $j$ and column $k$ of a reference image of size $M \times N$, and let $\widehat{F}(j,k)$ denote a testing image.

*Type I: IQMs based on pixel distance*

1. Average distance (AD):

$$AD = \sum_{j=1}^{M} \sum_{k=1}^{N} \left( F(j,k) - \widehat{F}(j,k) \right) / MN. \tag{11.1}$$

2. L2 distance (L2D):

$$L2D = \frac{1}{MN} \left( \sum_{j=1}^{M} \sum_{k=1}^{N} \left( F(j,k) - \widehat{F}(j,k) \right)^2 \right)^{1/2}. \tag{11.2}$$

*Type II: IQMs based on correlation*

3. Structure content (SC):

$$SC = \sum_{j=1}^{M} \sum_{k=1}^{N} F(j,k)^2 / \sum_{j=1}^{M} \sum_{k=1}^{N} \widehat{F}(j,k)^2. \tag{11.3}$$

4. Image fidelity (IF):

$$IF = 1 - \left( \sum_{j=1}^{M} \sum_{k=1}^{N} \left( F(j,k) - \widehat{F}(j,k) \right)^2 / \sum_{j=1}^{M} \sum_{k=1}^{N} \widehat{F}(j,k)^2 \right). \tag{11.4}$$

5. N cross-correlation (NK):

$$NK = \sum_{j=1}^{M} \sum_{k=1}^{N} F(j,k)\widehat{F}(j,k) / \sum_{j=1}^{M} \sum_{k=1}^{N} F(j,k)^2. \tag{11.5}$$

*Type III: IQMs based on mean square error*

6. Normal mean square error (NMSE):

$$NMSE = \sum_{j=1}^{M} \sum_{k=1}^{N} \left( F(j,k) - \widehat{F}(j,k) \right)^2 / \sum_{j=1}^{M} \sum_{k=1}^{N} F(j,k)^2. \tag{11.6}$$

7. Least mean square error (LMSE):

$$\text{LMSE} = \sum_{j=1}^{M-1}\sum_{k=2}^{N-1}(F(j,k)-\hat{F}(j,k))^2 \Big/ \sum_{j=1}^{M-1}\sum_{k=2}^{N-1}O(F(j,k))^2. \tag{11.7}$$

where $O(F(j,k)) = F(j+1,k)+\hat{F}(j-1,k)+F(j,k+1)+F(j,k-1)-4F(j,k)$

8. Peak mean square error (PMSE):

$$PMSE = \frac{1}{MN}\sum_{j=1}^{M}\sum_{k=1}^{N}[F(j,k)-\hat{F}(j,k)]^2 / \{\max_{j,k}[F(j,k)]\}^2. \tag{11.8}$$

9. Peak signal-to-noise ratio (PSNR):

$$PSNR = 20\times\log_{10}\left\{255\Big/\left\{\sum_{j=1}^{M}\sum_{k=1}^{N}[F(j,k)-\hat{F}(j,k)]^2\right\}^{1/2}\right\}. \tag{11.9}$$

Let $f(x_i)$ denote the IQM score of an image under the degree of distortion $x_i$. The IQMs used for measuring image blurs must satisfy the monotonically increasing or decreasing property. That is, if $x_{i+1} > x_i$, then $f(x_{i+1})-f(x_i) > 0$ or $f(x_{i+1})-f(x_i) < 0$. The sensitivity of the IQM is defined as the score of the aggregate relative distance, as follows:

$$\sum_{i=1}^{7}\frac{f(x_{i+1})-f(x_i)}{f(x_i)}. \tag{11.10}$$

A quality measure called the *mean opinion score* (MOS) was proposed by Grgic et al. [30] to correlate with various IQMs. However, it requires human interaction and is not suited for automatic processes. To measure the degree of "closeness" to the HVS, we develop the following procedure for measuring the HVS, where Nill's band-pass filter [31] is utilized.

1. Perform the DCT on the original image $C(x,y)$ and the distorted image $\hat{C}(x,y)$ to obtain $D(u,v)$ and $\hat{D}(u,v)$, respectively.
2. Convolve $D(u,v)$ and $\hat{D}(u,v)$ with the following band-pass filter to obtain $E(u,v)$ and $\hat{E}(u,v)$, respectively.

$$H(p) = \begin{cases} 0.05e^{p^{0.554}}, & \text{if } p < 7 \\ e^{-9|\log_{10}p-\log_{10}9|^{2.3}}, & \text{if } p \geq 7 \end{cases} \tag{11.11}$$

where $p = (u^2 + v^2)^{1/2}$.

3. Perform the inverse DCT of $E(u,v)$ and $\hat{E}(u,v)$ to obtain $F(x,y)$ and $\hat{F}(x,y)$, respectively.

4. Perform Euclidean distance computation on $F(x,y)$ and $\hat{F}(x,y)$ as follows:

$$d = \left\{ \sum_x \sum_y [F(x,y) - \hat{F}(x,y)]^2 \right\}^{1/2} .\tag{11.12}$$

### 11.2.4 LEARNING STRATEGY

There are many types of learning strategies, including statistical learning, neural networks, and *support vector machines* (SVMs). An SVM is intended to generate an optimal separating hyperplane by minimizing the generalization error without using the assumption of class probabilities such as Bayesian classifiers [32–35]. The decision hyperplane of an SVM is determined by the most informative data instances, called *support vectors* (SVs). In practice, these SVs are a subset of the entire training data. By applying feature reduction in both input and feature space, a fast nonlinear SVM can be designed without a noticeable loss in performance [36,37].

The training procedure of SVMs usually requires huge memory space and significant computation time due to the enormous amounts of training data and the quadratic programming problem. Some researchers have proposed incremental training or active learning to shorten the training time [38–43]. The goal is to select a subset of training samples while preserving the performance of using all the samples. Instead of learning from randomly selected samples, the actively learning SVM queries the informative samples in each incremental step.

Syed et al. [39] developed an incremental learning procedure by partitioning the training data set into subsets. At each incremental step, only SVs are added to the training set at the next step. Campbell et al. [40] proposed a query-learning strategy by iteratively requesting the label of a sample that is the closest to the current hyperplane. Schohn and Cohn [30] presented a greedy optimal strategy by calculating class probability and expected error. A simple heuristic is used for selecting training samples according to their proximity with respect to the dividing hyperplane.

Moreover, An et al. [41] developed an incremental learning algorithm by extracting initial training samples as the margin vectors of two classes, selecting new samples according to their distances to the current hyperplane and discarding the training samples that do not contribute to the separating plane. Nguyen and Smeulders [42] pointed out that prior data distribution is useful for active learning, and the new training data selected from the samples have maximal contribution to the current expected error.

Most existing methods of active learning measure proximity to the separating hyperplane. Mitra et al. [43] presented probabilistic active learning strategy using a distribution determined by the current separating hyperplane and confidence factor. Unfortunately, they did not provide the criteria for choosing the test samples. There are two issues in their algorithm. First, the initial training samples are selected randomly, which may generate the initial hyperplane to be far away from the optimal solution. Second, using only the SVs from the previous training set may lead to bad performance.

### 11.2.4.1    Introduction to Support Vector Machines

Cortes et al. [44] introduced two-group SVMs into classification. SVMs were originally designed as binary (or two-class) classifiers. Researchers have proposed extensions to multiclass (or multigroup) classification by combining multiple SVMs, such as *one-against-all*, *one-against-one*, and *directed acyclic graph* (DAG) SVMs [45]. The one-against-all method is to learn $k$ binary SVMs during the training stage. For the $i$th SVM, we label all training samples $y(j)_{j=i} = +1$ and $y(j)_{j \neq i} = -1$. In the testing, the decision function [46] is obtained by

$$\text{Class of } \bar{\mathbf{x}} = \arg \max_{i=1,2,\dots,k} ((w^i)^T \varphi(\bar{\mathbf{x}}) + b^i), \tag{11.13}$$

where $(w^i)^T \varphi(\bar{\mathbf{x}}) + b^i = 0$ represents the hyperplane for the $i$th SVM. It was shown that we can solve the $k$-group problem simultaneously [47].

The one-against-one method is to construct $k(k-1)/2$ SVMs in a tree structure. Each SVM represents a distinguishable pairwise classifier from different classes in the training stage. In the testing, each leaf SVM pops up one desirable class label, which propagates to the upper level in the tree until it processes the root SVM of the tree.

The DAG SVM [45] combines multiple binary one-against-one SVM classifiers into one multiclass classifier. It has $k(k-1)/2$ internal SVM nodes, but in the testing phase it starts from the root node and moves to the left or right subtree depending on the output value of the binary decision function until a leaf node is reached, which indicates the recognized class. There are two main differences between the one-against-one and DAG methods. One is that one-against-one needs to evaluate all $k(k-1)/2$ nodes, while DAG only needs to evaluate $k$ nodes due to the different testing phase schema. The other is that one-against-one uses a bottom-up approach, whereas DAG uses a top-down approach.

Suppose we have $l$ patterns, and each pattern consists of a pair: a vector $\mathbf{x}_i \in \mathbf{R}^n$ and the associated label $y_i \in \{-1,1\}$. Let $X \subset R^n$ be the space of the patterns, $Y = \{-1,1\}$ be the space of labels, and $p(\mathbf{x},y)$ be a probability density function of $X \times Y$. The machine-learning problem consists of finding a mapping, $sign(f(\mathbf{x}_i)) \rightarrow y_i$, which minimizes the error of misclassification. Let $f$ denote the decision function. The *sign* function is defined as

$$sign(f(\mathbf{x})) = \begin{cases} 1 & f(\mathbf{x}) \geq 0 \\ -1 & \text{otherwise.} \end{cases} \tag{11.14}$$

The expected value of the actual error is given by

$$E[e] = \int \frac{1}{2} |y - sign(f(\mathbf{x}))| p(\mathbf{x},y) d\mathbf{x}. \tag{11.15}$$

However, $p(\mathbf{x},y)$ is unknown in practice. We use the training data to derive the function $f$, whose performance is controlled by two factors: the training error and the capacity measured by the Vapnik–Chervonenkis (VC) dimension [35]. Let $0 \leq \eta \leq 1$. The following bound holds with a probability of $1 - \eta$.

$$e \leq e_t + \sqrt{\left(\frac{h(\log(2l/h)+1) - \log(\eta/4)}{l}\right)}, \tag{11.16}$$

where $h$ is the VC dimension and $e_t$ is the error tested on a training set as:

$$e_t = \frac{1}{2l} \sum_{i=1}^{l} |y_i - sign(f(\mathbf{x}_i))|. \tag{11.17}$$

For two-class classification, SVMs use a hyperplane that maximizes the margin (i.e., the distance between the hyperplane and the nearest sample of each class). This hyperplane is viewed as the *optimal separating hyperplane* (OSH). The training patterns are linearly separable if there exists a vector $\mathbf{w}$ and a scalar $b$, such that the following inequality is valid.

$$y_i(\mathbf{w}.\mathbf{x}_i + b) \geq 1, \; i = 1, 2, ..., l. \tag{11.18}$$

The margin between the two classes can be simply calculated as

$$m = \frac{2}{|\mathbf{w}|}. \tag{11.19}$$

Therefore, the OSH can be obtained by minimizing $|\mathbf{w}|^2$ subject to the constraint in Equation 11.18.

The positive Lagrange multipliers $\alpha_i \geq 0$ are introduced to form the Lagrangian as

$$L = \frac{1}{2}|\mathbf{w}|^2 - \sum_{i=1}^{l} \alpha_i y_i(\mathbf{x}_i \cdot \mathbf{w} + b) + \sum_{i=1}^{l} \alpha_i. \tag{11.20}$$

The optimal solution is determined by the saddle point of this Lagrangian, where the minimum is taken with respect to $\mathbf{w}$ and $b$, and the maximum is taken with respect to the Lagrange multipliers $\alpha_i$. By taking the derivatives of $L$ with respect to $\mathbf{w}$ and $b$, we obtain

$$\mathbf{w} = \sum_{i=1}^{l} \alpha_i y_i \mathbf{x}_i, \tag{11.21}$$

and

$$\sum_{i=1}^{l} \alpha_i y_i = 0. \tag{11.22}$$

Substituting Equations 11.21 and 11.22 into Equation 11.20, we obtain

$$L = \sum_{i=1}^{l} \alpha_i - \frac{1}{2} \sum_{i=1}^{l} \sum_{j=1}^{l} \alpha_i \alpha_j y_i y_j \mathbf{x}_i \cdot \mathbf{x}_j. \tag{11.23}$$

Thus, constructing the optimal hyperplane for the linear SVM is a quadratic programming problem that maximizes Equation 11.23, subject to Equation 11.22.

If the data are not linearly separable in the input space, a nonlinear transformation function $\Phi(\cdot)$ is used to project $\mathbf{x}_i$ from the input space $\mathbf{R}^n$ to a higher dimensional feature space $\mathbf{F}$. The OSH is constructed in the feature space by maximizing the margin between the closest points $\Phi(\mathbf{x}_i)$. The inner product between two projections is defined by a kernel function $K(\mathbf{x}, \mathbf{y}) = \Phi(\mathbf{x}) \cdot \Phi(\mathbf{y})$. The commonly used kernels include the polynomial, Gaussian *radial basis function* (RBF), and Sigmoid kernels. Similar to linear SVM, constructing the OSH for the SVM with kernel $K(\mathbf{x}, \mathbf{y})$ involves maximizing

$$L = \sum_{i=1}^{l} \alpha_i - \frac{1}{2} \sum_{i=1}^{l} \sum_{j=1}^{l} \alpha_i \alpha_j y_i y_j K(\mathbf{x}_i, \mathbf{x}_j), \tag{11.24}$$

subject to Equation 11.22.

In practical applications, the training data in two classes are not often completely separable. In this case, a hyperplane that maximizes the margin while minimizing the number of errors is desirable. It is obtained by maximizing Equation 11.24, subject to the constraints $\sum_{i=1}^{l} \alpha_i y_i = 0$ and $0 \le \alpha_i \le C$, where $\alpha_i$ is the Lagrange multiplier for pattern $\mathbf{x}_i$ and $C$ is the regularization constant that manages the trade-off between the minimization of the number of errors and the necessity to control the capacity of the classifier.

### 11.2.4.2 Neural Networks

A layered neural network is a network of neurons organized in the form of layers. The first layer is the input layer. In the middle, there are one or more hidden layers. The last one is the output layer. The architecture of a layered neural network is shown in Figure 11.1. The function of the hidden neurons is to intervene between the external input and the network output. The output vector $\vec{\mathbf{Y}}(\vec{\mathbf{x}})$ is

$$\vec{\mathbf{Y}}(\vec{\mathbf{x}}) = \sum_{i=1}^{l} f_i(\vec{\mathbf{x}}) w_{ik} = \vec{\mathbf{F}} \cdot W, \tag{11.25}$$

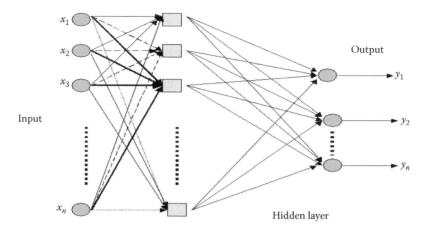

**FIGURE 11.1**   Architecture of a layered neural network.

where:

$\vec{F}$   is the hidden layer's output vector corresponding to a sample $\vec{x}$

$W$   is the weight matrix with dimensions of $l \times k$

$l$   is the number of hidden neurons and $k$ is the number of output nodes

In such a structure, all neurons cooperate to generate a single output vector. In other words, the output not only depends on each individual neuron but on all of them.

There are two categories in neural networks. One is supervised learning and the other is unsupervised learning. For supervised learning, the target responses are known for the training set, but for unsupervised learning, we do not know the desired responses. In this chapter, we use supervised learning.

### 11.2.4.3   Principal Component Analysis

*Principal component analysis* (PCA) is known as the most suitable data representation in a least-square sense for classical recognition techniques [48]. It can reduce the dimensionality of original images dramatically and retain most information. Suppose we have a population of $n$ faces as the training set $\vec{z} = (\vec{z}_1, \vec{z}_2, ... \vec{z}_n)^T$, where each component $\vec{z}_i$ represents a training image vector. The mean vector of this population is computed by

$$\vec{\mu} = \frac{1}{N} \sum_{i=1}^{N} \vec{z}_i.$$   (11.26)

The covariance matrix is computed by

$$S = \frac{1}{N-1} \sum_{i=1}^{N} (\vec{z}_i - \vec{\mu})(\vec{z}_i - \vec{\mu})^T.$$   (11.27)

Using the covariance matrix, we calculate the eigenvectors and eigenvalues. Based on nonzero eigenvalues in a decreasing order, we use the corresponding eigenvectors to construct the eigenspace $U$ (also known as the *feature space*) as $U = (\bar{\mathbf{e}}_1, \bar{\mathbf{e}}_2, ..., \bar{\mathbf{e}}_k)$, where $\bar{\mathbf{e}}_j$ is the $j$th eigenvector. Then, we project all images onto this eigenspace to find new representation for each image. The new representation is computed by

$$\bar{\mathbf{x}}_i = U^T(\bar{\mathbf{z}}_i - \bar{\boldsymbol{\mu}}). \tag{11.28}$$

Through the PCA process, we can reduce original images from a high-dimensional input space to a much lower-dimensional feature space. Note that the PCA transformation can preserve the structure of original distribution without degrading classification.

### 11.2.5  FREQUENCY DOMAIN STEGANALYTIC SYSTEM

Similar to watermarking, steganography can embed hidden messages into the frequency domain by using its redundancies. The simplest method is LSB substitution. There are various methods of changing the frequency coefficients for reducing the statistical artifacts in the embedding process—for example, the *F5* [49] and *OutGuess* [50] algorithms.

Pfitzmann and Westfeld [49] developed the F5 algorithm to achieve robust and high-capacity JPEG steganography. They embed messages by modifying the DCT coefficients. The F5 algorithm has two important features; first, it permutes the DCT coefficients before embedding, and second, it employs matrix embedding. The first operation—namely, permuting the DCT coefficients—has the effect of spreading the changed coefficients evenly over the entire image. The second operation is matrix embedding. They use the nonzero non-DC coefficients and the message length to calculate a suitable matrix embedding that will minimize the number of changes in the cover image. Although the algorithm alters the histogram of the DCT coefficients, they have verified that some characteristics, such as the monotonicity of the increments and of the histogram, are in fact preserved.

Fridrich et al. [51] presented a steganalytic method that detects hidden messages in JPEG images using the F5 algorithm. It estimates the histogram of the cover image from the stego-image. This can be accomplished by decompressing the stego-image, cropping the image by four pixels in both directions, and recompressing the stego-image by the same quality factor.

OutGuess is a steganographic system available in UNIX. The first version, OutGuess 0.13b, is vulnerable to statistical analysis. The second version, OutGuess 0.2, incorporates the preservation of statistical properties [50]. OutGuess embeds the message into the selected DCT coefficients by a pseudorandom number generator. A pseudorandom number generator and a stream cipher are created by a user-defined pass phrase. Basically, there are two steps in OutGuess to embed the message. It first detects the redundant DCT coefficients with the least effect on the cover image. Then it selects the bits in which to embed the message according to the obtained information.

Recent newspapers stories indicate that terrorists may communicate secret information to each other using steganography, hiding messages in images on the Internet. In order to analyze whether there are steganographic messages on the Internet, Provos and Honeyman [17] presented several tools to retrieve and perform automatic analysis on Internet images. Alturki and Mersereau [52] proposed quantizing the coefficients in the frequency domain to embed secret messages. They first decorrelate the image by scrambling the pixels randomly, which in effect whitens the frequency domain of the image and increases the number of transform coefficients in the frequency domain, thus increasing the embedding capacity. The frequency coefficients are then quantized to even or odd multiples of the quantization step size to embed 1s or 0s. The inverse Fourier transform is then taken and descrambled. The resulting image is visually incomparable to the original image.

## REFERENCES

1. Wrixon, F. B., *Codes, Ciphers and Other Cryptic and Clandestine Communication*, Black Dog & Leventhal, New York, 469–508, 1998.
2. Simmons, G. J., Prisoners' problem and the subliminal channel, in *Proc. Int. Conf. Advances in Cryptology*, 1984, 51.
3. Fei, C., Kundur, D., and Kwong, R. H., Analysis and design of secure watermark-based authentication systems, *IEEE Trans. Information Forensics and Security*, 1, 43, 2006.
4. Lyu, S. and Farid, H., Steganalysis using higher-order image statistics, *IEEE Trans. Information Forensics and Security*, 1, 111, 2006.
5. Sullivan, K. et al., Steganalysis for Markov cover data with applications to images, *IEEE Trans. Information Forensics and Security*, 1, 275, 2006.
6. Johnson, N. and Jajodia, S., Exploring steganography: Seeing the unseen, *IEEE Computer*, 31, 26, 1998.
7. Johnson, N. F., Durric, Z., and Jajodia, S., *Information Hiding: Steganography and Watermarking*, Kulwer Academic, Boston, MA, 2001.
8. Kharrazi, M., Sencar, H. T., and Memon, N. D., Benchmarking steganographic and steganalysis techniques, in *Proc. Int. Conf. Security, Steganography, and Watermarking of Multimedia Contents*, San Jose, CA, 2005.
9. Wu, Y. and Shih, F. Y., Genetic algorithm based methodology for breaking the steganalytic systems, *IEEE Trans. SMC: Part B*, 36, 24, 2006.
10. Brown, A., S-Tools for Windows, shareware, 1994.
11. Wolf, B., StegoDos: Black Wolf's Picture Encoder v0.90B, public domain, ftp://ftp.csua. berkeley.edu/pub/cypherpunks/steganography/stegodos.zip. Accessed January 6, 2017.
12. Korejwa, J., JSteg shell 2.0., http://www.tiac.net/users/korejwa/steg.htm. Accessed January 6, 2017.
13. Westfeld, A. and Pfitzmann, A., Attacks on steganographic systems breaking the steganographic utilities EzStego, Jsteg, Steganos, and S-Tools and some lessons learned, in *Proc. Int. Workshop Information Hiding*, Dresden, Germany, 1999, 61.
14. Avcibas, I., Memon, N., and Sankur, B., Steganalysis using image quality metrics, *IEEE Trans. Image Processing*, 12, 221, 2003.
15. Fridrich, J., Goljan, M., and Hogea, D., New methodology for breaking steganographic techniques for JPEGs, in *Proc. EI SPIE*, Santa Clara, CA, 2003, 143.
16. Farid, H., Detecting steganographic message in digital images, Technical Report, TR2001–412, Department of Computer Science, Dartmouth College, Hanover, NH, 2001.

17. Provos, N. and Honeyman, P., Detecting steganographic content on the Internet, Technical Report 01–11, Center for Information Technology Integration, University of Michigan, Ann Arbor, 2001.
18. Fridrich, J. et al., Quantitative steganalysis of digital images: Estimating the secret message length, *Multimedia Systems*, 9, 288, 2003.
19. Wayner, P., *Disappearing Cryptography: Information Hiding; Steganography and Watermarking*, 2nd edn., Morgan Kaufmann, San Francisco, CA, 2002.
20. Jackson, J. T. et al., Blind steganography detection using a computational immune system: A work in progress, *Int. J. Digital Evidence*, 4, 19, 2003.
21. Fridrich, J., Goljan, M., and Du, R., Reliable detection of LSB steganography in color and grayscale images, in *Proc. ACM Workshop Multimedia Security*, Ottawa, ON, Canada, 2001, 27.
22. Westfeld, A. and Pfitzmann, A., Attacks on steganographic systems, *Lecture Notes in Computer Science*, vol. 1768, 61, Springer, Berlin, 2000.
23. Fridrich, J., Goljan, M., Soukal, D., Higher-order statistical steganalysis of palette images, in *Proc. SPIE Conf. Security and Watermarking of Multimedia Contents V*, 2003, 178.
24. Dumitrescu, S., Wu, X., and Wang, Z., Detection of LSB steganography via sample pair analysis, in *Proc. Information Hiding Workshop*, 2002, 355.
25. Fridrich, J. and Du, R., Secure steganographic methods for palette images, in *Proc. Information Hiding Workshop*, Dresden, Germany, 1999, 4760.
26. Fridrich, J. and Goljan, M., Practical steganalysis of digital images: State of the art, in *Proc. SPIE Security and Watermarking of Multimedia Contents IV*, San Jose, CA, 2002, 1.
27. Ozer, H. et al., Steganalysis of audio based on audio quality metrics, in *Proc. SPIE, Security and Watermarking of Multimedia Contents V*, Santa Clara, CA, 2003, 55.
28. Kessler, G. C., An overview of steganography for the computer forensics examiner, *Forensic Science Communications*, 6, 2004.
29. Avcibas, I., Memon, N., and Sankur, B., Steganalysis using image quality metrics, *IEEE Trans. Image Processing*, 12, 221, 2003.
30. Grgic, S., Grgic, M., and Mrak, M., Reliability of objective picture quality measures, *J. Electrical Engineering*, 55, 3, 2004.
31. Nill, N. B., A visual model weighted cosine transform for image compression and quality assessment, *IEEE Trans. Communication*, 33, 551, 1985.
32. Cortes, C. and Vapnik, V., Support-vector networks, *Machine Learning*, 20, 273, 1995.
33. Vapnik, V., *The Nature of Statistical Learning Theory*, Springer, New York, 1995.
34. Burges, C., A tutorial on support vector machines for pattern recognition, *IEEE Trans. Data Mining and Knowledge Discovery*, 2, 121, 1998.
35. Vapnik, V., *Statistical Learning Theory*, Wiley, New York, 1998.
36. Heisele, B. et al., Hierarchical classification and feature reduction for fast face detection with support vector machines, *Pattern Recognition*, 36, 2007, 2003.
37. Shih, F. Y. and Cheng, S., Improved feature reduction in input and feature spaces, *Pattern Recognition*, 38, 651, 2005.
38. Schohn, G. and Cohn, D., Less is more: Active learning with support vector machines, in *Proc. Int. Conf. Machine Learning*, Stanford University, CA, 2000, 839.
39. Syed, N. A., Liu, H., and Sung, K. K., Incremental learning with support vector machines, in *Proc. Int. Joint Conf. Artificial Intelligence*, Stockholm, Sweden, 1999, 15.
40. Campbell, C., Cristianini, N., and Smola, A., Query learning with large margin classifiers, in *Proc. Int. Conf. Machine Learning*, Stanford University, CA, 2000, 111.

41. An, J.-L., Wang, Z.-O., and Ma, Z.-P., An incremental learning algorithm for support vector machine, in *Proc. Int. Conf. Machine Learning and Cybernetics*, Xi'an, China, 2003, 1153.

42. Nguyen, H. T. and Smeulders, A., Active learning using pre-clustering, in *Proc. Int. Conf. Machine Learning*, Banff, AB, Canada, 2004.

43. Mitra, P., Murthy, C. A., and Pal, S. K., A probabilistic active support vector learning algorithm, *IEEE Trans. Pattern Analysis and Machine Intelligence*, 26, 413, 2004.

44. Cortes, C. and Vapnik, V., Support-vector network, *Machine Learning*, 20, 273, 1995.

45. Platt, J. C., Cristianini, N., and Shawe-Taylor, J., Large margin DAGs for multiclass classification, in *Advances in Neural Information Processing Systems*, MIT Press, 12, 547, 2000.

46. Hsu, C. W. and Lin, C. J., A comparison of methods for multiclass support vector machines, *IEEE Trans. Neural Networks*, 13, 415, 2002.

47. Herbrich, R., *Learning Kernel Classifiers Theory and Algorithms*, MIT Press, Cambridge, MA, 2001.

48. Martinez, A. M. and Kak, A. C., PCA versus LDA, *IEEE Trans. Pattern Analysis and Machine Intelligence*, 23, 228, 2001.

49. Westfeld, A., High capacity despite better steganalysis (F5—a steganographic algorithm), in *Proc. Int. Workshop Information Hiding*, Pittsburgh, PA, 2001, 289.

50. Provos, N., Defending against statistical steganalysis, in *Proc. USENIX Security Symposium*, 2001, 323.

51. Fridrich, J., Goljan, M., and Hogea, D., Steganalysis of JPEG images: Breaking the F5 algorithm, in *Proc. Int. Workshop Information Hiding*, Noordwijkerhout, The Netherlands, 2002, 310.

52. Alturki, F. and Mersereau, R., A novel approach for increasing security and data embedding capacity in images for data hiding applications, in *Proc. Int. Conf. Information Technology: Coding and Computing*, 2001, 228.

# 12 Steganography Based on Genetic Algorithms and Differential Evolution

Steganography aims at hiding information into covert channels, so that one can conceal the information and prevent the detection of hidden messages. For steganographic systems, the fundamental requirement is that the stego-object (the object containing hidden messages) should be perceptually indistinguishable to the degree that it does not raise suspicion. In other words, the hidden information introduces only slight modifications to the cover object (the original object containing no hidden messages). For image steganography, most passive wardens distinguish the stego-images by analyzing their statistical features. Since steganalytic systems analyze certain statistical features of an image, the idea of developing a robust steganographic system is to generate the stego-image while avoiding changes to the statistical features of the cover image.

Chu et al. [1] presented a DCT-based steganographic system that utilizes the similarities between the DCT coefficients of the adjacent image blocks where the embedding distortion is spread. Their algorithm randomly selects DCT coefficients in order to maintain key statistical features. However, the drawback of their approach is that the capacity of the embedded message is limited—that is, only 2 bits for an $8 \times 8$ DCT block. In this chapter, we present the techniques of breaking steganalytic systems based on *genetic algorithms* (GAs) as well as *differential evolution* (DE) [2,3]. The emphasis is shifted from traditionally avoiding the alteration of statistical features to artificially counterfeiting them. Our idea is based on the following: in order to manipulate the statistical features, the GA- or DE-based approach is adopted to counterfeit several stego-images (*candidates*) until one of them can break the inspection of the steganalytic systems.

Due to the fact that GAs converge slowly to the optimal solution, other evolutionary approaches have been developed, such as DE and *evolution strategy* (ES) [4]. One drawback of ES is that it is easier to get stuck on a local optimum. We use DE to increase the performance of the steganographic system because it is a simple and efficient adaptive scheme for global optimization over continuous space. It can find the global optimum of a multidimensional, multimodal function with good probability.

This chapter is organized as follows: Section 12.1 presents steganography based on GAs; Section 12.2 describes steganography based on DE.

## 12.1   STEGANOGRAPHY BASED ON GENETIC ALGORITHMS

We first describe an overview of the GA-based breaking methodology in Section 12.1.1. We then present GA-based breaking algorithms in *spatial domain steganalytic systems* (SDSSs) and *frequency domain steganalytic systems* (FDSSs) in Sections 12.1.2 and 12.1.3, respectively. Experimental results are shown in Section 12.1.4. Complexity analysis is provided in Section 12.1.5.

### 12.1.1   OVERVIEW OF THE GA-BASED BREAKING METHODOLOGY

The GA methodology, introduced by Holland [5] in his seminal work, is commonly used as an adaptive approach that provides a randomized, parallel, and global search based on the mechanics of natural selection and genetics in order to find solutions to problems. In general, a GA starts with some randomly selected genes as the first generation, called the *population*. Each individual in the population corresponding to a solution in the problem domain is called a *chromosome*. An objective, called the *fitness function*, is used to evaluate the quality of each chromosome. The higher-quality chromosomes will survive and form the new population of the next generation. We can recombine a new generation to find the best solution using three operators: *reproduction*, *crossover*, and *mutation*. The process is repeated until a predefined condition is satisfied or a constant number of iterations is reached; we can then correctly extract the desired hidden message.

In order to apply a GA for embedding messages into the frequency domain of a cover image to obtain a stego-image, we use the chromosome $\xi$, consisting of $n$ genes ($\xi = g_0, g_1, g_2, \ldots, g_n$). Figure 12.1 gives an example of a chromosome ($\xi \in Z^{64}$) containing 64 genes ($g_i \in Z[\text{integers}]$), where (a) shows the distribution order of a chromosome in an $8 \times 8$ block and (b) shows an example of the corresponding chromosome.

The chromosome is used to adjust the pixel values of the cover image to generate a stego-image, so the embedded message can be correctly extracted and at the same time the statistical features can be retained in order to break the steganalytic system. A fitness function is used to evaluate the embedded message and statistical features.

| $g_0$ | $g_1$ | $g_2$ | $g_3$ | $g_4$ | $g_5$ | $g_6$ | $g_7$ |
|---|---|---|---|---|---|---|---|
| $g_8$ | $g_9$ | $g_{10}$ | $g_{11}$ | $g_{12}$ | $g_{13}$ | $g_{14}$ | $g_{15}$ |
| $g_{16}$ | $g_{17}$ | $g_{18}$ | $g_{19}$ | $g_{20}$ | $g_{21}$ | $g_{22}$ | $g_{23}$ |
| $g_{24}$ | $g_{25}$ | $g_{26}$ | $g_{27}$ | $g_{28}$ | $g_{29}$ | $g_{30}$ | $g_{31}$ |
| $g_{32}$ | $g_{33}$ | $g_{34}$ | $g_{35}$ | $g_{36}$ | $g_{37}$ | $g_{38}$ | $g_{39}$ |
| $g_{40}$ | $g_{41}$ | $g_{42}$ | $g_{43}$ | $g_{44}$ | $g_{45}$ | $g_{46}$ | $g_{47}$ |
| $g_{48}$ | $g_{49}$ | $g_{50}$ | $g_{51}$ | $g_{52}$ | $g_{53}$ | $g_{54}$ | $g_{55}$ |
| $g_{56}$ | $g_{57}$ | $g_{58}$ | $g_{59}$ | $g_{60}$ | $g_{61}$ | $g_{62}$ | $g_{63}$ |

(a)

| 0 | 0 | 1 | 0 | 0 | 2 | 0 | 1 |
|---|---|---|---|---|---|---|---|
| 1 | 2 | 0 | 0 | 1 | 1 | 0 | 1 |
| 2 | 1 | 1 | 0 | 0 | -2 | 0 | 1 |
| 3 | 1 | 2 | 5 | 0 | 0 | 0 | 0 |
| 0 | 2 | 0 | 0 | 0 | 1 | 0 | 0 |
| 0 | -1 | 0 | 2 | 1 | 1 | 4 | 2 |
| 1 | 1 | 1 | 0 | 0 | 0 | 0 | 0 |
| 0 | 0 | 0 | 0 | 1 | -2 | 1 | 3 |

(b)

**FIGURE 12.1**   The numbering positions corresponding to 64 genes.

Let $C$ and $S$ denote the cover image and the stego-image of size $8 \times 8$, respectively. We generate the stego-image by adding together the cover image and the chromosome as follows:

$$S = \left\{ s_i \mid s_i = c_i + g_i \right\}, \tag{12.1}$$

where $0 \le i \le 63$.

#### 12.1.1.1 Fitness Function

In order to embed messages into DCT-based coefficients and avoid the detection of steganalytic systems, we develop a fitness function to evaluate the following two terms:

1. *Analysis*($\xi$,$C$): The analysis function evaluates the difference between the cover image and the stego-image in order to maintain the statistical features. It is related to the type of steganalytic system used and will be explained in Sections 12.1.2 and 12.1.3.
2. *BER*($\xi$,$C$): The *bit error rate* (BER) sums up the bit differences between the embedded and the extracted messages. It is defined as

$$BER(\xi,C) = \frac{1}{\mid Message^H \mid} \sum_{i=0}^{all\ pixels} \mid Message_i^H - Message_i^E \mid, \tag{12.2}$$

where:
$Message^H$ is the embedded binary message
$Message^E$ is the extracted binary message
$\mid Message^H \mid$ is the length of the message

For example, if $Message^H = 11111$ and $Message^E = 10101$, then $BER(\xi,C) = 0.4$.

We use a linear combination of the analysis and the bit error rate to be the fitness function, as follows:

$$Evaluation\ (\xi,C) = \alpha_1 \times Analysis(\xi,C) + \alpha_2 \times BER(\xi,C), \tag{12.3}$$

where $\alpha_1$ and $\alpha_2$ denote the weights. The weights can be adjusted according to the user's demand for the degree of distortion to the stego-image or the extracted message.

#### 12.1.1.2 Reproduction

$$Reproduction\left(\Psi,k\right) = \left\{ \xi_i \mid Evaluation(\xi_i,C) \le \Omega \ \text{for} \ \xi_i \in \Psi \right\}, \tag{12.4}$$

where $\Omega$ is a threshold for sieving chromosomes and $\Psi = \{\xi_1, \xi_2, ..., \xi_n\}$. It is used to reproduce $k$ higher-quality chromosomes from the original population.

#### 12.1.1.3 Crossover

$$Crossover(\Psi,l) = \left\{ \xi_i \ominus \xi_j \mid \xi_i, \xi_j \in \Psi \right\}, \tag{12.5}$$

where $\Theta$ denotes the operation to generate chromosomes by exchanging genes from their parents, $\xi_i$ and $\xi_j$. It is used to gestate $l$ higher-quality offspring by inheriting better genes (i.e., those found to be of a higher quality in the fitness evaluation) from their parents. One-point, two-point, and multipoint crossovers are often used. The criteria of selecting a suitable crossover depend on the length and structure of the chromosomes. We adopt the one- and two-point crossovers, as shown in [6].

### 12.1.1.4 Mutation

$$Mutation(\Psi, m) = \left\{ \xi_i \circ j \mid 0 \le j \le |\xi_i| \text{ and } \xi_i \in \Psi \right\}, \tag{12.6}$$

where $\circ$ denotes the operation to randomly select a chromosome $\xi_i$ from $\Psi$ and change the $j$th bit from $\xi_i$. It is used to generate $m$ new chromosomes. The mutation is usually performed with a probability $p$ $(0 < p \le 1)$, meaning only $p$ of the genes in a chromosome will be selected to be mutated. Since the length of a chromosome is 64 bits and there are only one or two genes to be mutated when the GA mutation operation is performed, we select $p$ to be 1/64 or 2/64.

Note that in each generation, the new population is generated by the preceding three operations. The new population is actually the same size $(k + l + m)$ as the original population. Note that, to break the inspection of the steganalytic systems, we use a straightforward GA selection method whereby the new population is generated based on the chromosomes having superior evaluation values to the previous generation. For example, only the top 10% of the current chromosomes will be considered for reproduction, crossover, and mutation.

The mutation tends to be more efficient than the crossover if a candidate solution is close to the real optimum solution. In order to enhance the performance of our GA-based methodology, we generate a new chromosome with the desired minor adjustments when the previous generation is close to the goal. Therefore, the strategy of dynamically determining the ratio of the three GA operations is utilized. Let $R_P$, $R_M$, and $R_C$ denote the ratios of production, mutation, and crossover, respectively. In the beginning, we select a small $R_P$ and $R_M$ and a large $R_C$ to enable the global search. After a certain number of iterations, we will decrease $R_C$ and increase $R_M$ and $R_P$ to shift focus onto a more local search if the current generation is better than the old one; otherwise, we will increase $R_C$ and decrease $R_M$ and $R_P$ to enlarge the range of the global search. Note that $R_P + R_M + R_C = 100\%$ must be satisfied.

The recombination strategy of our GA-based algorithm is presented in the following section. We apply the same strategy to recombine the chromosome, except that the fitness function is defined differently with respect to the properties of the individual problem.

### 12.1.1.5 Algorithm for Recombining Chromosomes

1. Initialize the base population of chromosomes, $R_P$, $R_M$, and $R_C$.
2. Generate candidates by adjusting the pixel values of the original image.
3. Determine the fitness value of each chromosome.
4. If a predefined condition is satisfied or a constant number of iterations is reached, the algorithm will stop and output the best chromosome to

be the solution; otherwise, go to the following steps to recombine the chromosomes.

5. If a certain number of iterations is reached, go to step 6 to adjust $R_P$, $R_M$, and $R_C$; otherwise, go to step 7 to recombine the new chromosomes.
6. If 20% of the new generation are better than the best chromosome in the preceding generation, then $R_C = R_C - 10\%$, $R_M = R_M + 5\%$, and $R_P = R_P + 5\%$.
7. Otherwise, $R_C = R_C + 10\%$, $R_M = R_M - 5\%$, and $R_P = R_P - 5\%$. Obtain the new generation by recombining the preceding chromosomes using production, mutation, and crossover.
8. Go to Step 2.

## 12.1.2   GA-Based Breaking Algorithms in Spatial Domain Steganalytic Systems

In order to generate a stego-image that passes through the inspection of an SDSS, the messages should be embedded into specific positions to maintain the statistical features of the cover image. In the spatial domain embedding approach, it is difficult to select such positions since the messages are distributed regularly. On the other hand, if the messages are embedded into the specific positions of the coefficients of a transformed image, the changes in the spatial domain are difficult to predict. We intend to find the desired positions in the frequency domain that produce minimal statistical feature disturbance in the spatial domain. We apply a GA to generate a stego-image by adjusting the pixel values in the spatial domain using the following two criteria:

1. Evaluate the extracted messages obtained from the specific coefficients of a stego-image to be as close as possible to the embedded messages.
2. Evaluate the statistical features of the stego-image and compare them with those of the cover image such that there are as few differences as possible.

### 12.1.2.1   Generating Stego-images in the Visual Steganalytic System

Westfeld and Pfitzmann [7] presented a *visual steganalytic system* (VSS) that uses an assignment function of color replacement called a *visual filter* to efficiently detect a stego-image by translating the grayscale image into binary. Therefore, in order to break the VSS, the two results ($VF^C$ and $VF^S$) of applying a visual filter to the cover images and stego-images, respectively, should be as identical as possible. The VSS was originally designed to detect GIF-format images by reassigning the color in the color palette of an image. We extend this method to detect BMP-format images as well by setting the odd- and even-numbered gray scales to black and white, respectively. The $Analysis(\xi, C)$ of the VSS, indicating the sum of difference between $LSB^C$ and $LSB^S$, is defined as

$$Analysis(\xi, C) = \frac{1}{|C|} \sum_{i=0}^{all\_pixels} \left( VF_i^C \oplus VF_i^S \right), \tag{12.7}$$

where $\oplus$ denotes the *exclusive or* (XOR) operator. Our algorithm is described as follows:

*Algorithm for generating a stego-image for the VSS*:

1. Divide a cover image into a set of cover images of size $8 \times 8$.
2. For each $8 \times 8$ cover image, generate a stego-image based on a GA to perform the embedding procedure as well as to ensure that $LSB^C$ and $LSB^S$ are as identical as possible.
3. Combine all the $8 \times 8$ stego-images together to form a complete stego-image.

### 12.1.2.2 Generating Stego-images in the Image Quality Measure–Based Spatial Domain Steganalytic System

Avcibas et al. [8] proposed a steganalytic system that analyzes the *image quality measures* (IQMs) [9] of the cover image and the stego-images. The *IQM-based spatial domain steganalytic system* (IQM-SDSS) consists of two phases: training and testing. In the training phase, the IQMs are calculated between an image and the image filtered using a low-pass filter based on the Gaussian kernel. Suppose there are $N$ images and $q$ IQMs in the training set. Let $x_{ij}$ denote the score in the $i$th image and the $j$th IQM, where $1 \leq i \leq N$ and $1 \leq i \leq q$. Let $y_i$ be the value of $-1$ or $1$, indicating the cover images or stego-image, respectively. We can represent all the images in the training set as

$$
\begin{aligned}
y_1 &= \beta_1 x_{11} + \beta_2 x_{12} + \cdots + \beta_q x_{1q} + \varepsilon_1 \\
y_2 &= \beta_1 x_{21} + \beta_2 x_{22} + \cdots + \beta_q x_{2q} + \varepsilon_2 \\
&\vdots \\
y_N &= \beta_1 x_{N1} + \beta_2 x_{N2} + \cdots + \beta_q x_{Nq} + \varepsilon_N
\end{aligned}
\qquad (12.8)
$$

where $\varepsilon_1, \varepsilon_2, \ldots, \varepsilon_N$ denote random errors in the linear regression model [10]. Then, the linear predictor $\beta = [\beta_1, \beta_2, \ldots, \beta_q]$ can be obtained from all the training images.

In the testing phase, we use the $q$ IQMs to compute $y_i$ to determine whether it is a stego-image. If $y_i$ is positive, the test image is a stego-image; otherwise, it is a cover image.

We first train the IQM-SDSS in order to obtain the linear predictor—that is, $\beta = [\beta_1, \beta_2, \ldots, \beta_q]$—from our database. Then, we use $\beta$ to generate the stego-image with our GA-based algorithm, so that it can pass through the inspection of the IQM-SDSS. Note that the GA procedure is not used in the training phase. The *Analysis*$(\xi, C)$ of the IQM-SDSS is defined as

$$
Analysis(\xi, C) = \beta_1 x_1 + \beta_2 x_2 + \cdots + \beta_q x_q. \qquad (12.9)
$$

*Algorithm for generating a stego-image for the IQM-SDSS*:

1. Divide a cover image into a set of cover images of size $8 \times 8$.
2. Adjust the pixel values in each $8 \times 8$ cover image based on the GA to embed messages into the frequency domain and ensure that the stego-image can

pass through the inspection of the IQM-SDSS. The procedures for generating the $8 \times 8$ stego-image are presented after this algorithm.

3. Combine all the $8 \times 8$ embedded images together to form a completely embedded image.
4. Test the embedded image with the IQM-SDSS. If it passes, it is the desired stego-image; otherwise, repeat steps 2 to 4.

*Procedure for generating an $8 \times 8$ embedded image for the IQM-SDSS*:

1. Define the fitness function, the number of genes, the size of population, the crossover rate, the critical value, and the mutation rate.
2. Generate the first generation by random selection.
3. Generate an $8 \times 8$ embedded image based on each chromosome.
4. Evaluate the fitness value of each chromosome by analyzing the $8 \times 8$ embedded image.
5. Obtain the best chromosome based on the fitness values.
6. Recombine new chromosomes by crossover.
7. Recombine new chromosomes by mutation.
8. Repeat steps 3 to 8 until a predefined condition is satisfied or a constant number of iterations is reached.

### 12.1.3 GA-BASED BREAKING ALGORITHMS IN FREQUENCY DOMAIN STEGANALYTIC SYSTEMS

Fridrich et al. [11] presented a steganalytic system for detecting JPEG stego-images based on the assumption that the histogram distributions of specific *alternating current* (AC) DCT coefficients of a cover image and its cropped image should be similar. Note that, in DCT coefficients, the zero frequency (0,0) is the only *direct current* (DC) component, and the remaining frequencies are AC components. Let $h_{kl}(d)$ and $\bar{h}_{kl}(d)$ denote the total number of AC DCT coefficients in the $8 \times 8$ cover image and its corresponding $8 \times 8$ cropped images, respectively, with the absolute value equal to $d$ at location $(k,l)$, where $0 \le k$ and $l \le 7$. Note that the $8 \times 8$ cropped images defined in Fridrich et al. [11] are obtained in the same way as the $8 \times 8$ cover images with a horizontal shift by 4 pixels.

The probability $\rho_{kl}$ of the modification of a nonzero AC coefficient at $(k,l)$ can be obtained by

$$\rho_{kl} = \frac{\bar{h}_{kl}(1)\left[h_{kl}(0) - \bar{h}_{kl}(0)\right] + \left[h_{kl}(1) - \bar{h}_{kl}(1)\right]\left[\bar{h}_{kl}(2) - \bar{h}_{kl}(1)\right]}{\left[\bar{h}_{kl}(1)\right]^2 + \left[\bar{h}_{kl}(2) - \bar{h}_{kl}(1)\right]^2}. \quad (12.10)$$

Note that the final value of the parameter $\rho$ is calculated as an average of the selected low-frequency DCT coefficients—that is, $(k,l) \in \{(1,2), (2,1), (2,2)\}$—and only 0, 1, and 2 are considered when checking the coefficient values of the specific frequencies between a cover image and its corresponding cropped image.

Our GA-based breaking algorithm for the *JPEG frequency domain steganalytic system* (JFDSS) is intended to minimize the differences between the two histograms of a stego-image and its cropped image. It is presented as follows:

*Algorithm for generating a stego-image for the JFDSS*:

1. Compress a cover image by JPEG and divide it into a set of small cover images of size $8 \times 8$. Each is performed by the DCT.
2. Embed the messages into the specific DCT coefficients and decompress the embedded image by the IDCT.
3. Select a $12 \times 8$ working window and generate an $8 \times 8$ cropped image for each $8 \times 8$ embedded image.
4. Determine the overlapping area between each $8 \times 8$ embedded image and its cropped image.
5. Adjust the overlapping pixel values, making sure that the coefficients of specific frequencies $(k,l)$ of the stego-image and its cropped image as identical as possible and that the embedded messages are not altered.
6. Repeat steps 3 to 6 until all the $8 \times 8$ embedded images are generated.

Let $Coef^{Stego}$ and $Coef^{Crop}$ denote the coefficients of each $8 \times 8$ stego-image and its cropped image. The $Analysis(\xi,C)$ of the JFDSS is defined as

$$Analysis(\xi,C) = \frac{1}{|C|} \sum_{i=0}^{all\ pixels} (Coef_i^{Stego} \otimes Coef_i^{Crop}), \qquad (12.11)$$

where $\otimes$ denotes the operator defined by

$$\begin{cases} Coef_i^{Stego} \otimes Coef_i^{Crop} = 1 & \text{if } (Coef_i^{Stego} = 0 \text{ and } Coef_i^{Crop} \neq 0) \text{ or} \\ & (Coef_i^{Stego} = 1 \text{ and } Coef_i^{Crop} \neq 1) \text{ or} \\ & (Coef_i^{Stego} = 2 \text{ and } Coef_i^{Crop} \neq 2) \text{ or} \\ & (Coef_i^{Stego} \neq 0,1,2 \text{ and } Coef_i^{Crop} = 0,1,2) \\ Coef_i^{Stego} \otimes Coef_i^{Crop} = 0 & \text{otherwise.} \end{cases} \qquad (12.12)$$

Note that 0, 1, and 2 denote the values in the specific frequencies obtained by dividing the quantization table. We only consider the values of the desired frequencies to be 0, 1, 2, or the values in Equation 12.12 because of the strategy of the JFDSS in Equation 12.10.

Figure 12.2 shows an example of our GA-based algorithm for the JFDSS. In Figure 12.2a, we select a $12 \times 8$ working window for each $8 \times 8$ stego-image and generate its $8 \times 8$ cropped image. Note that the shaded pixels indicate their overlapping area, and the three black boxes in Figure 12.2b are the desired locations.

## 12.1.4   EXPERIMENTAL RESULTS

In this section, we provide experimental results to show that our GA-based steganographic method can successfully break the inspection of steganalytic systems. To test our algorithm, we used a database of 200 grayscale images of size $256 \times 256$. All the images were originally stored in the BMP format.

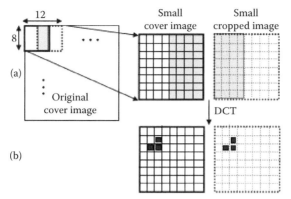

The frequency domain of cover image and cropped images

**FIGURE 12.2**   An example of our GA-based algorithm in the JFDSS.

### 12.1.4.1   GA-Based Breaking Algorithm in the Visual Steganalytic System

We test our algorithm in the VSS. Figures 12.3a,f show a stego-image and a message image of sizes $256 \times 256$ and $64 \times 64$, respectively. We embed 4 bits into the $8 \times 8$ DCT coefficients on frequencies (0,2), (1,1), (2,0), and (3,0) to avoid distortion. Note that the stego-image in Figure 12.3a is generated by embedding Figure 12.3f into the DCT coefficients of the cover image using our GA-based algorithm. Figures 12.3b–j, respectively, display the bit planes from 7 to 0. Figure 12.4 shows the stego-image and its visually filtered result. It is difficult to determine whether Figure 12.4a is a stego-image.

Figure 12.5 shows the relationship of the average iteration for adjusting an $8 \times 8$ cover image versus the correct rate of the visual filter. The correct rate is the percentage of similarity between the transformed results of the cover images and stego-images using the visual filter. Note that the BERs in Figure 12.5 are all 0%.

### 12.1.4.2   GA-Based Breaking Algorithm in the IQM-Based Spatial Domain Steganalytic System

We generate three stego-images as training samples for each cover image with Photoshop plug-in Digimarc [12], Cox's technique [13], and S-Tools [14]. Therefore, there are a total of 800 images of size $256 \times 256$, including 200 cover images and 600 stego-images. The embedded message sizes are 1/10, 1/24, and 1/40 of the cover image size for Digimarc, Cox's technique, and S-Tools, respectively. Note that the IQM-SDSS [15] can detect stego-images containing a message size of 1/100 of the cover image. We develop the following four training strategies to obtain the linear predictors.

1. Train all the images in the database to obtain the linear predictor $\beta^A$.
2. Train 100 cover images and 100 stego-images to obtain the linear predictor $\beta^B$, in which the stego-images are obtained by Cox's technique.
3. Train 100 cover images and 100 stego-images to obtain the linear predictor $\beta^C$, in which the stego-images are obtained by Digimarc.

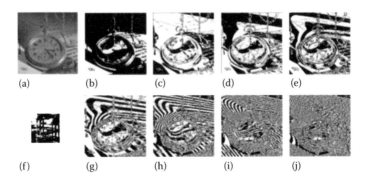

(a)          (b)          (c)          (d)          (e)

(f)          (g)          (h)          (i)          (j)

**FIGURE 12.3** A stego-image generated by our GA-based algorithm and its eight bit planes.

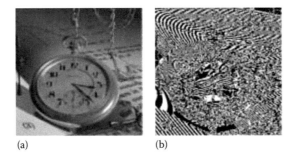

(a)                    (b)

**FIGURE 12.4** A stego-image and its visually filtered result.

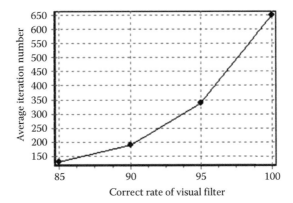

**FIGURE 12.5** The relationship of the average iteration vs. the correct rate of the visual filter.

4. Train 100 cover images and 100 stego-images to obtain the linear predictor $\beta^D$, in which the stego-images are obtained by S-tools.

In the testing phase, we generate 50 stego-images for each linear predictor. Therefore, we obtain four types of stego-images, $SI_A$, $SI_B$, $SI_C$, and $SI_D$, corresponding

**TABLE 12.1**

**Experimental Results of GA-Based Algorithm in the IQM-SDSS**

|  |  | $\beta^A$ | $\beta^B$ | $\beta^C$ | $\beta^D$ | Average |
|---|---|---|---|---|---|---|
| 50 $SI_A$ images | FN |  | 94% | 94% | 100% | 96% |
| 50 $SI_B$ images | FN | 84% |  | 84% | 80% | 83% |
| 50 $SI_C$ images | FN | 82% | 86% |  | 88% | 85% |
| 50 $SI_D$ images | FN | 82% | 86% | 82% |  | 83% |
| 50 cover images | FP | 6% | 10% | 16% | 16% | 12% |

to $\beta^A$, $\beta^B$, $\beta^C$, and $\beta^D$, respectively. It is obvious that all the $SI_A$ images will pass through the inspection of the IQM-SDSS with the linear predictor $\beta^A$ but may fail with other linear predictors—likewise for $SI_B$, $SI_C$, and $SI_D$.

Table 12.1 shows the false-negative (FN; Type I Error) and false-positive (FP; Type II Error) rates of testing each type of stego-image and cover image under different linear predictors. For example, the FN rate of testing 50 $SI_A$ stego-images under $\beta^B$ is 94%, indicating that the $\beta^B$ IQM-SDSS falsely decides that 94% of our stego-images belong to the cover image. On the other hand, the FP rate of testing 50 cover images under $\beta^B$ is 10%, indicating that the $\beta^B$ IQM-SDSS decides that 10% of the cover images belong to the stego-images. Note that the cover images used in the testing phase are different from those used in the training phase.

### 12.1.4.3   GA-Based Breaking Algorithm in the JPEG Frequency Domain Steganalytic System

Figure 12.6 shows an example of adjusting an $8 \times 8$ embedded image to obtain an $8 \times 8$ stego-image for breaking the JFDSS. Figure 12.6a shows a $12 \times 8$ working window, where the enclosed is the overlapping area. Figures 12.6b,c show the original $8 \times 8$ embedded and cropped images, respectively. We embed 1s on (1,2), (2,1), and (2,2) by compressing Figure 12.6b using JPEG under 70% compression quality to obtain Figure 12.6d. Note that the top-left pixel is (0,0) and the messages can be embedded into any frequency of the transformed domain, so that the embedding capacity will be sufficiently high. As the JFDSS only checks frequencies (1,2), (2,1), and (2,2), we show an example of embedding 3 bits into these three frequencies. Similarly, Figure 12.6e is obtained by compressing Figure 12.6c using JPEG under the same compression quality. By evaluating the frequencies (1,2), (2,1), and (2,2), the JFDSS can determine whether the embedded image is a stego-image. Therefore, in order to break the JFDSS, we obtain the stego-image in Figure 12.6f. Figures 12.6g and 12.6h, respectively, show the new embedded and cropped images. Similarly, we obtain Figures 12.6i and 12.6j by compressing Figures 12.6g and 12.6h, respectively, using JPEG under 70% compression quality. Therefore, the JFDSS cannot distinguish the frequencies (1,2), (2,1), and (2,2).

Let $QTable(i,j)$ denote the standard quantization table, where $0 \leq I$ and $j \leq 7$. The new quantization table, $NewTable(i,j)$, with $x\%$ compression quality can be obtained by

(a)

| 169 | 166 | 162 | 158 | 157 | 158 | 161 | 163 |
|-----|-----|-----|-----|-----|-----|-----|-----|
| 166 | 164 | 162 | 159 | 159 | 159 | 161 | 162 |
| 162 | 162 | 161 | 161 | 161 | 161 | 161 | 161 |
| 158 | 159 | 162 | 162 | 162 | 162 | 160 | 160 |
| 157 | 159 | 161 | 163 | 163 | 162 | 160 | 158 |
| 158 | 159 | 161 | 162 | 162 | 160 | 158 | 157 |
| 160 | 161 | 161 | 160 | 160 | 158 | 157 | 156 |
| 162 | 162 | 161 | 159 | 159 | 158 | 157 | 156 |

(b)

| 169 | 166 | 162 | 158 | 157 | 158 | 161 | 163 |
|-----|-----|-----|-----|-----|-----|-----|-----|
| 166 | 164 | 162 | 159 | 159 | 159 | 161 | 162 |
| 162 | 162 | 161 | 161 | 161 | 161 | 161 | 161 |
| 158 | 159 | 162 | 162 | 162 | 162 | 160 | 160 |
| 157 | 159 | 161 | 163 | 163 | 161 | 160 | 158 |
| 158 | 159 | 161 | 162 | 162 | 160 | 158 | 157 |
| 160 | 161 | 161 | 160 | 160 | 158 | 157 | 156 |
| 162 | 162 | 161 | 159 | 159 | 158 | 157 | 156 |

(c)

| 157 | 158 | 161 | 163 | 157 | 158 | 161 | 163 |
|-----|-----|-----|-----|-----|-----|-----|-----|
| 159 | 159 | 161 | 162 | 159 | 159 | 161 | 162 |
| 161 | 161 | 161 | 161 | 161 | 161 | 161 | 161 |
| 162 | 162 | 160 | 160 | 162 | 162 | 160 | 160 |
| 163 | 161 | 162 | 158 | 163 | 161 | 160 | 158 |
| 162 | 160 | 158 | 157 | 162 | 160 | 158 | 157 |
| 160 | 160 | 158 | 157 | 160 | 160 | 158 | 157 |
| 158 | 158 | 157 | 156 | 158 | 157 | 156 | 156 |

(d)

| 127 | 1 | 0 | 0 | 0 | 0 | 0 | 0 |
|-----|---|---|---|---|---|---|---|
| 1 | 0 | 1 | 0 | 0 | 0 | 0 | 0 |
| 0 | 1 | 1 | 0 | 0 | 0 | 0 | 0 |
| 0 | 0 | 0 | 0 | 0 | 0 | 0 | 0 |
| 0 | 0 | 0 | 0 | 0 | 0 | 0 | 0 |
| 0 | 0 | 0 | 0 | 0 | 0 | 0 | 0 |
| 0 | 0 | 0 | 0 | 0 | 0 | 0 | 0 |
| 0 | 0 | 0 | 0 | 0 | 0 | 0 | 0 |

(e)

| 126 | 1 | $-1$ | 0 | 0 | 0 | 0 | 0 |
|-----|---|------|---|---|---|---|---|
| 1 | $-1$ | 0 | 0 | 0 | 0 | 0 | 0 |
| $-1$ | 0 | 0 | 0 | 0 | 0 | 0 | 0 |
| 0 | 0 | 0 | 0 | 0 | 0 | 0 | 0 |
| 0 | 0 | 0 | 0 | 0 | 0 | 0 | 0 |
| 0 | 0 | 0 | 0 | 0 | 0 | 0 | 0 |
| 0 | 0 | 0 | 0 | 0 | 0 | 0 | 0 |
| 0 | 0 | 0 | 0 | 0 | 0 | 0 | 0 |

(f)

| 169 | 166 | 162 | 158 | 164 | 163 | 164 | 159 | 164 | 158 | 161 | 163 |
|-----|-----|-----|-----|-----|-----|-----|-----|-----|-----|-----|-----|
| 166 | 164 | 162 | 159 | 162 | 162 | 162 | 161 | 162 | 159 | 161 | 162 |
| 162 | 162 | 161 | 161 | 162 | 162 | 161 | 161 | 161 | 161 | 161 | 161 |
| 158 | 159 | 162 | 162 | 160 | 160 | 160 | 161 | 161 | 162 | 160 | 160 |
| 157 | 159 | 161 | 163 | 158 | 160 | 160 | 159 | 160 | 163 | 160 | 158 |
| 158 | 159 | 161 | 162 | 162 | 160 | 159 | 158 | 159 | 162 | 158 | 157 |
| 160 | 161 | 161 | 160 | 160 | 159 | 161 | 158 | 158 | 160 | 157 | 156 |
| 162 | 162 | 161 | 159 | 159 | 161 | 161 | 159 | 159 | 159 | 156 | 156 |

(g)

| 169 | 166 | 162 | 158 | 164 | 163 | 164 | 159 |
|-----|-----|-----|-----|-----|-----|-----|-----|
| 166 | 164 | 162 | 159 | 162 | 162 | 162 | 161 |
| 162 | 162 | 161 | 161 | 162 | 162 | 161 | 161 |
| 158 | 159 | 162 | 162 | 160 | 160 | 159 | 160 |
| 157 | 159 | 161 | 162 | 160 | 160 | 157 | 159 |
| 158 | 159 | 161 | 162 | 160 | 160 | 159 | 158 |
| 160 | 161 | 161 | 161 | 159 | 161 | 161 | 158 |
| 162 | 162 | 161 | 162 | 161 | 159 | 158 | 159 |

(h)

| 164 | 163 | 164 | 159 | 157 | 158 | 161 | 163 |
|-----|-----|-----|-----|-----|-----|-----|-----|
| 162 | 162 | 162 | 161 | 159 | 159 | 161 | 162 |
| 162 | 162 | 161 | 161 | 161 | 161 | 161 | 161 |
| 160 | 162 | 161 | 161 | 162 | 160 | 160 | 160 |
| 160 | 158 | 159 | 160 | 163 | 161 | 160 | 158 |
| 160 | 160 | 157 | 159 | 162 | 160 | 158 | 157 |
| 161 | 159 | 161 | 158 | 158 | 159 | 157 | 156 |
| 161 | 161 | 158 | 157 | 156 | 156 | 156 | 156 |

(i)

| 127 | 1 | 1 | 0 | 0 | 0 | 0 | 0 |
|-----|---|---|---|---|---|---|---|
| 1 | 0 | 1 | 0 | 0 | 0 | 0 | 0 |
| 0 | 1 | 1 | 0 | 0 | 0 | 0 | 0 |
| 0 | 0 | 0 | 0 | 0 | 0 | 0 | 0 |
| 0 | 0 | 0 | 0 | 0 | 0 | 0 | 0 |
| 0 | 0 | 0 | 0 | 0 | 0 | 0 | 0 |
| 0 | 0 | 0 | 0 | 0 | 0 | 0 | 0 |
| 0 | 0 | 0 | 0 | 0 | 0 | 0 | 0 |

(j)

| 127 | 1 | 1 | 0 | 0 | 0 | 0 | 0 |
|-----|---|---|---|---|---|---|---|
| 1 | 0 | 1 | 0 | 0 | 0 | 0 | 0 |
| 0 | 1 | 1 | 0 | 0 | 0 | 0 | 0 |
| 0 | 0 | 0 | 0 | 0 | 0 | 0 | 0 |
| 0 | 0 | 0 | 0 | 0 | 0 | 0 | 0 |
| 0 | 0 | 0 | 0 | 0 | 0 | 0 | 0 |
| 0 | 0 | 0 | 0 | 0 | 0 | 0 | 0 |
| 0 | 0 | 0 | 0 | 0 | 0 | 0 | 0 |

**FIGURE 12.6**   An example of adjusting an 8 × 8 embedded image.

| 16 | 11 | 10 | 16 | 24 | 40 | 51 | 61 |
|----|----|----|----|----|-----|-----|----|
| 12 | 12 | 14 | 19 | 26 | 58 | 60 | 55 |
| 14 | 13 | 16 | 24 | 40 | 57 | 69 | 56 |
| 14 | 17 | 22 | 29 | 51 | 87 | 80 | 62 |
| 18 | 22 | 37 | 56 | 68 | 109 | 103 | 77 |
| 24 | 35 | 55 | 64 | 81 | 104 | 113 | 92 |
| 49 | 64 | 78 | 87 | 103 | 121 | 120 | 101 |
| 72 | 92 | 95 | 98 | 112 | 100 | 103 | 99 |

(a)

| 10.1 | 7.1 | 6.5 | 10.1 | 14.9 | 24.5 | 31.1 | 37.1 |
|------|-----|-----|------|------|------|------|------|
| 7.7 | 7.7 | 8.9 | 11.9 | 16.1 | 35.3 | 36.5 | 33.5 |
| 8.9 | 8.3 | 10.1 | 14.9 | 24.5 | 34.7 | 41.9 | 34.1 |
| 8.9 | 10.7 | 13.7 | 17.9 | 31.1 | 52.7 | 48.5 | 37.7 |
| 11.3 | 13.7 | 22.7 | 34.1 | 41.3 | 65.9 | 62.3 | 46.7 |
| 14.9 | 21.5 | 33.5 | 38.9 | 49.1 | 62.9 | 68.3 | 55.7 |
| 29.9 | 38.9 | 47.3 | 52.7 | 62.3 | 73.1 | 72.5 | 61.1 |
| 43.7 | 55.7 | 57.5 | 59.3 | 67.7 | 60.5 | 62.3 | 59.9 |

(b)

**FIGURE 12.7**   The JPEG quantization table.

$$NewTable(i, j) = \frac{QTable(i, j) \times factor + 50}{100}, \qquad (12.13)$$

where the factor is determined by

$$\begin{cases} factor = \dfrac{5000}{x} & \text{if } x \leq 50 \\ factor = 200 - 2x & \text{otherwise.} \end{cases} \qquad (12.14)$$

Figures 12.7a and 12.7b show the quantization tables at standard quality and 70% compression, respectively.

### 12.1.5   COMPLEXITY ANALYSIS

In general, the complexity of our GA-based algorithm is related to the size of the embedded message and the position of embedding [15]. Figure 12.8 shows the relationship between the embedded message and the required iterations, in which the cover image is of size $8 \times 8$ and the message is embedded into the LSB of the cover image by the zigzag order starting from the DC (zero-frequency) component. We observe that the more messages embedded into a stego-image, the more iterations are required in our GA-based algorithm. Figure 12.9 shows an example where we embed a 4-bit message into the different bit planes of DCT coefficients. We observe that the lower the bit plane used for embedding, the more iterations are required in our GA-based algorithm.

## 12.2   STEGANOGRAPHY BASED ON DIFFERENTIAL EVOLUTION

This section is organized as follows: the DE algorithm and its image-based steganography [16] are presented in Section 12.2.1; experimental results are shown in Section 12.2.2.

### 12.2.1   DE ALGORITHM AND ITS IMAGE-BASED STEGANOGRAPHY

DE is a kind of *evolutionary algorithm* (EA), which derives its name from natural biological evolutionary processes. EAs are mimicked to obtain a solution for an

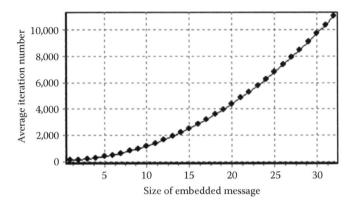

**FIGURE 12.8**    The relationship between the size of the embedded message and the required iterations.

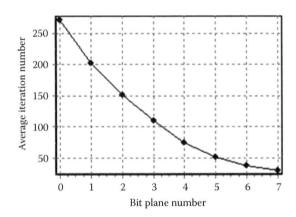

**FIGURE 12.9**    Embedding a 4-bit message into a different bit plane.

optimization problem. One can use EAs for problems that are difficult to solve by traditional optimization techniques, including the problems that are not well defined or are difficult to model mathematically. A popular optimization method that belongs to the EA class of methods is the GA. In recent years, GAs have replaced traditional methods to be the preferred optimization tools, as several studies have conclusively proved. DE [3] is a relative latecomer, but its popularity has been catching up. It is fast in numerical optimization and is more likely to find the true optimum.

### 12.2.1.1    Concept

The method of DE mainly consists of four steps: initialization, mutation, recombination, and selection. In the first step, a random population of potential solutions is created within the multidimensional search space. To start with, we define the objective function $f(\mathbf{y})$ to be optimized, where $\mathbf{y} = (y^1, y^2, \ldots, y^n)$ is a vector of $n$ decision variables. The aim is to find a vector $\mathbf{y}$ in the given search space, for which the value of the objective function is an optimum. The search space is first determined by providing the lower and upper bounds for each of the $n$ decision variables of $\mathbf{y}$.

In the initialization step, a population of $NP$ vectors, each of $n$ dimensions, is randomly created. The parameters are encoded as floating-point numbers. Mutation is basically a search mechanism that directs the search toward potential areas of optimal solution together with recombination and selection. In this step, three distinct target vectors $\mathbf{y}_a$, $\mathbf{y}_b$, and $\mathbf{y}_c$ are randomly chosen from the $NP$ parent population on the basis of three random numbers $a$, $b$, and $c$. One of the vectors, $\mathbf{y}_c$, is the base of the mutated vector. This is added to the weighted difference of the remaining two vectors—that is, $\mathbf{y}_a - \mathbf{y}_b$—to generate a noisy random vector, $\mathbf{n}_i$. The weighting is done using scaling factor $F$, which is user supplied and usually in the range 0 to 1.2. This mutation process is repeated to create a mate for each member of the parent population.

In the recombination (crossover) operation, each target vector of the parent population is allowed to undergo the recombination by mating with a mutated vector. Thus, vector $\mathbf{y}_i$ is recombined with the noisy random vector $\mathbf{n}_i$ to generate a trial vector, $\mathbf{t}_i$. Each element of the trial vector ($t_i{}^m$, where $i = 1, 2, \ldots, NP$ and $m = 1, 2, \ldots, n$) is determined by a binomial experiment whose success or failure is determined by the user-supplied crossover factor $CR$. The parameter $CR$ is used to control the rate at which the crossover takes place. The trial vector, $\mathbf{t}_i$, is thus the child of two parent vectors: noisy random vector $\mathbf{n}_i$ and target vector $\mathbf{y}_i$. DE performs a nonuniform crossover, which determines the trial vector parameters to be inherited from which parent.

It is possible that particular combinations of three target vectors from the parent population and the scaling factor $F$ will result in noisy vector values that are outside the bounds set for the decision variables. It is necessary, therefore, to bring such values within the bounds. For this reason, the value of each element of the trial vector is checked at the end of the recombination step. If it violates the bounds, it is heuristically brought back to lie within the bounded region. In the last stage of the selection, the trial vector is pitted against the target vector. The fitness is tested, and the fitter of the two vectors survives and proceeds to the next generation.

After $NP$ competitions of this kind in each generation, one will have a new population that is fitter than that of the previous generation. This evolutionary procedure, consisting of the preceding four steps, is repeated over several generations until the termination condition is satisfied (i.e., when the objective function attains a prescribed optimum) or a specified number of generations is reached, whichever happens earlier. The flowchart is described in Figure 12.10.

### 12.2.1.2   Algorithm

1. The first step is the random initialization of the parent population. Each element of $NP$ vectors in $n$ dimensions is randomly generated as

$$y_i^m = y_{min}^m + rand(0,1) \times (y_{max}^m - y_{min}^m), \tag{12.15}$$

where $i = 1, 2, \ldots, NP$ and $m = 1, 2, \ldots, n$.

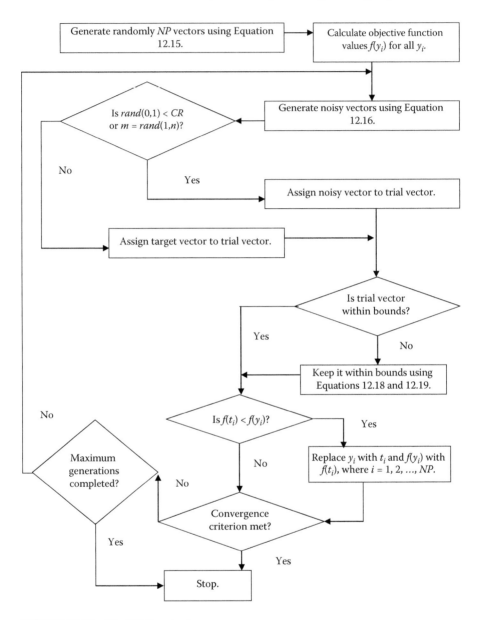

**FIGURE 12.10**    The DE flowchart.

2. Calculate the objective function values $f(\mathbf{y}_i)$ for all $\mathbf{y}_i$.
3. Select three random numbers ($a$, $b$, and $c$) within the range 1 to $NP$. The weighted difference $(\mathbf{y}_a - \mathbf{y}_b)$ is used to perturb $\mathbf{y}_c$ to generate a noisy vector $\mathbf{n}_i$.

$$\mathbf{n}_i = \mathbf{y}_c + F \times (\mathbf{y}_a - \mathbf{y}_b),$$  (12.16)

where $i = 1, 2, \ldots, NP$.

4. Recombine each target vector $\mathbf{y}_i$ with the noisy random vector $\mathbf{n}_i$ to generate a trial vector $\mathbf{t}_i$.

$$\begin{cases} t_i^m = n_i^m & \text{if } rand(0,1) < CR \text{ or } m = rand(1,n) \\ t_i^m = y_i^m & \text{otherwise,} \end{cases} \tag{12.17}$$

where $i = 1, 2, \ldots, NP$ and $m = 1, 2, \ldots, n$.

Check whether each decision variable of the trail vector is within the bounds. If it is outside the bonds, force it to lie within the bounds using the following:

$$t_i^m = y_{\min}^m + 2 \times (p/q) \times (y_{\max}^m - y_{\min}^m), \tag{12.18}$$

where:

$$\begin{cases} p = t_i^m - y_{\max}^m \text{ and } q = t_i^m - y_{\min}^m & \text{if } t_i^m > y_{\max}^m \\ p = y_{\min}^m - t_i^m \text{ and } q = y_{\max}^m - t_i^m & \text{otherwise.} \end{cases} \tag{12.19}$$

6. Calculate the value of the objective function for the two vectors $\mathbf{t}_i$ and $\mathbf{y}_i$. The fitter of the two (i.e., the one with the lower objective function value) survives and proceeds to the next generation.
7. Check if the convergence criterion meets. If yes, stop; otherwise, go to step 8.
8. Check if the maximum number of generations is reached. If yes, stop; otherwise, go to step 3.

### 12.2.1.3    Procedure of Applying Differential Evolution

Because the DE algorithm deals with floating-point numbers, the procedure of generating the stego-image is slightly different from the GA algorithm. The evaluation function is the same as that of the GA procedure. It uses the following rule to generate a stego-image: if $d_{ij} \geq 0.5$, $s_{ij} = c_{ij} + 1$; otherwise, $s_{ij} = c_{ij}$, where $1 \leq I$ and $j \leq 8$, and $s_{ij}$ and $c_{ij}$ denote the pixels in the stego-image and cover image, respectively. This procedure is illustrated in Figure 12.11, where (a) is the generated $8 \times 8$ DE vector by the DE algorithm, (b) is the $8 \times 8$ cover image, and (c) is the corresponding stego-image.

### 12.2.2    Experimental Results

The GA- and DE-based methodologies for breaking the VSS was implemented in Java. The Java Distributed Evolutionary Algorithms Library (JDEAL) was used to implement the GA. Experiments were conducted on 50 cover images of size $256 \times 256$ and 10 secret messages of size $64 \times 64$. Some results are tabulated in Table 12.2 for iterations of 25, 50, and 100.

From observing Table 12.2, it is clear that for the same number of iterations, the DE algorithm not only improves the PSNR values of the stego-image but also promotes the *normalized correlation* (NC) values of the secret messages. The percentage

| 0.43 | 0.23 | 0.36 | 0.78 | 0.82 | 0.46 | 0.78 | 0.65 |
|------|------|------|------|------|------|------|------|
| 0.98 | 0.14 | 0.09 | 0.36 | 0.49 | 0.61 | 0.48 | 0.24 |
| 0.96 | 0.84 | 0.98 | 0.92 | 0.87 | 0.88 | 0.94 | 0.64 |
| 0.14 | 0.19 | 0.03 | 0.78 | 0.86 | 0.94 | 0.42 | 0.56 |
| 0.64 | 0.54 | 0.13 | 0.26 | 0.37 | 0.48 | 0.87 | 0.64 |
| 0.94 | 0.64 | 0.44 | 0.24 | 0.92 | 0.51 | 0.66 | 0.16 |
| 0.24 | 0.45 | 0.56 | 0.64 | 0.46 | 0.24 | 0.76 | 0.43 |
| 0.24 | 0.94 | 0.48 | 0.65 | 0.78 | 0.92 | 0.93 | 0.46 |

(a)

| 12  | 67  | 98  | 245 | 129 | 67  | 230 | 11  |
|-----|-----|-----|-----|-----|-----|-----|-----|
| 119 | 142 | 231 | 12  | 65  | 98  | 114 | 167 |
| 24  | 46  | 78  | 134 | 145 | 64  | 25  | 142 |
| 98  | 211 | 198 | 46  | 98  | 34  | 12  | 9   |
| 97  | 74  | 96  | 43  | 123 | 213 | 56  | 78  |
| 87  | 46  | 215 | 87  | 13  | 58  | 34  | 156 |
| 98  | 46  | 16  | 178 | 126 | 214 | 101 | 98  |
| 19  | 36  | 47  | 58  | 36  | 24  | 126 | 48  |

(b)

| 12  | 67  | 98  | 246 | 130 | 67  | 231 | 12  |
|-----|-----|-----|-----|-----|-----|-----|-----|
| 120 | 142 | 231 | 12  | 65  | 99  | 114 | 167 |
| 25  | 47  | 79  | 135 | 146 | 65  | 26  | 143 |
| 98  | 211 | 198 | 47  | 99  | 35  | 12  | 10  |
| 98  | 75  | 96  | 43  | 123 | 213 | 57  | 79  |
| 88  | 47  | 215 | 87  | 14  | 59  | 35  | 156 |
| 98  | 46  | 17  | 179 | 126 | 214 | 102 | 98  |
| 19  | 37  | 47  | 59  | 37  | 25  | 127 | 48  |

(c)

**FIGURE 12.11**    (a) DE vector, (b) cover image, (c) stego-image.

**TABLE 12.2**

**Comparison of Image Steganography Using GA and DE Algorithms**

| Image | Watermark | Iteration | GA PSNR | GA NC | DE PSNR | DE NC |
|-------|-----------|-----------|---------|-------|---------|-------|
| *Lena* | Chinese character | 25 | 52.122 | 0.973 | 54.037 | 0.995 |
| | | 50 | 52.588 | 0.979 | 55.502 | 1.0 |
| | | 100 | 53.146 | 0.993 | 58.160 | 1.0 |
| *Barbara* | Chinese character | 25 | 51.133 | 0.967 | 53.963 | 0.999 |
| | | 50 | 51.140 | 0.967 | 55.385 | 0.999 |
| | | 100 | 51.146 | 0.993 | 57.972 | 1.0 |
| *Cameraman* | Chinese character | 25 | 51.143 | 0.918 | 54.048 | 0.976 |
| | | 50 | 51.132 | 0.955 | 55.553 | 0.979 |
| | | 100 | 51.151 | 0.974 | 58.130 | 0.987 |
| *Lena* | Flower | 25 | 51.125 | 0.8993 | 53.883 | 0.985 |
| | | 50 | 51.133 | 0.965 | 55.228 | 0.993 |
| | | 100 | 51.148 | 0.981 | 57.659 | 0.9926 |
| *Barbara* | Flower | 25 | 51.135 | 0.968 | 53.949 | 0.993 |
| | | 50 | 51.134 | 0.985 | 55.317 | 0.997 |
| | | 100 | 51.134 | 0.991 | 57.804 | 1.0 |
| *Cameraman* | Flower | 25 | 51.150 | 0.937 | 53.942 | 0.959 |
| | | 50 | 51.138 | 0.933 | 55.282 | 0.985 |
| | | 100 | 51.151 | 0.961 | 57.549 | 0.990 |

(b)

(a)

**FIGURE 12.12**   Images before and after the application of DE methodology to break the VSS. (a) Stego-image after 25 iterations of each block using DE, (b) secret message retrieved from (a).

increase in PSNR values ranges from 5% to 13%, and that of NC values ranges from 0.8% to 3%.

Figure 12.12 shows the results obtained by embedding a $64 \times 64$ Chinese character into a $256 \times 256$ *Lena* image for 25 iterations of each $8 \times 8$ block using DE. Figure 12.13 shows the results obtained by embedding a $64 \times 64$ flower image into a $256 \times 256$ *Barbara* image for 100 iterations of each $8 \times 8$ block using a GA and DE. Figure 12.14 shows the results obtained by embedding a $64 \times 64$ Chinese character into a $256 \times 256$ *Cameraman* image for 50 iterations of each $8 \times 8$ block using a GA and DE. By comparing Figures 12.13d and 12.13f, and Figures 12.14d and 12.14f, we can observe that the extracted secret messages from the DE-generated stego-images are of a higher quality than those from the GA-generated stego-images.

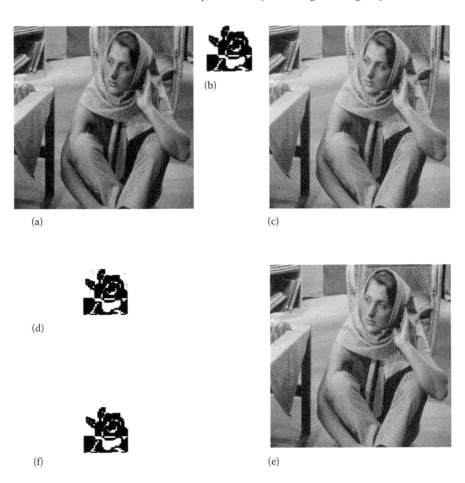

**FIGURE 12.13**    The images before and after applying the GA and DE algorithms: (a) *Barbara* $256 \times 256$ cover image, (b) Flower $64 \times 64$ watermark, (c) *Barbara* stego-image after 100 iterations using a GA, (d) watermark extracted from (c), (e) *Barbara* stego-image after 100 iterations using DE, (f) watermark extracted from (e).

**FIGURE 12.14** The images before and after applying the GA and DE algorithms: (a) Cameraman $256 \times 256$ cover image, (b) Chinese character $64 \times 64$ watermark, (c) Cameraman stego-image after 50 iterations using GA, (d) Extracted watermark from (c), (e) Cameraman stego-image after 50 iterations using DE, (f) Extracted watermark from (e).

## REFERENCES

1. Chu, R. et al., A DCT-based image steganographic method resisting statistical attacks, in *Int. Conf. Acoustics, Speech, and Signal Processing*, Montreal, QC, Canada, 2004.
2. Karboga, D. and Okdem, S., A simple and global optimization algorithm for engineering problems: Differential evolution algorithm, *Turkish J. Electrical Engineering and Computer Sciences*, 12, 53, 2004.
3. Price, K., Storn, R., and Lampinen, J., *Differential Evolution: A Practical Approach to Global Optimization*, Springer, New York, 2005.
4. Beyer, H. and Schwefel, H., Evolution strategies: A comprehensive introduction, *Natural Computing*, 1, 3, 2004.
5. Holland, J. H., *Adaptation in Natural and Artificial Systems*, University of Michigan Press, Ann Arbor, 1975.

6. Shih, F. Y. and Wu, Y.-T., Enhancement of image watermark retrieval based on genetic algorithm, *J. Visual Communication and Image Representation*, 16, 115, 2005.

7. Westfeld, A. and Pfitzmann, A., Attacks on steganographic systems breaking the steganographic utilities EzStego, Jsteg, Steganos, and S-Tools and some lessons learned, in *Proc. Int. Workshop Information Hiding*, Dresden, Germany, 1999, 61.

8. Avcibas, I., Memon, N., and Sankur, B., Steganalysis using image quality metrics, *IEEE Trans. Image Processing*, 12, 221, 2003.

9. Avcibas, I. and Sankur, B., Statistical analysis of image quality measures, *J. Electronic Imaging*, 11, 206, 2002.

10. Rencher, A. C., *Methods of Multivariate Analysis*, John Wiley, New York, 1995.

11. Fridrich, J., Goljan, M., and Hogea, D., New methodology for breaking steganographic techniques for JPEGs, in *Proc. EI SPIE*, Santa Clara, CA, 2003, 143.

12. PictureMarc, Embed Watermark, v 1.00.45, Digimarc Corporation.

13. Cox, J. et al., Secure spread spectrum watermarking for multimedia, *IEEE Trans. Image Processing*, 6, 1673, 1997.

14. Brown, A., S-Tools for Windows, shareware, 1994. ftp://idea.sec.dsi.unimi.it/pub/security/crypt/code/s-tools4.zip. Accessed January 6, 2017.

15. Shih, F. Y. and Wu, Y.-T., Combinational image watermarking in the spatial and frequency domains, *Pattern Recognition*, 36, 969, 2003.

16. Shih, F. Y. and Edupuganti, V. G., A differential evolution based algorithm for breaking the visual steganalytic system, *Soft Computing*, 13, 345, 2009.

# Index